exam success

in

ACCOUNTING

for Cambridge International AS & A Level

David Austen

OXFORD
UNIVERSITY PRESS

Great Clarendon Street, Oxford, OX2 6DP, United Kingdom

Oxford University Press is a department of the University of Oxford.
It furthers the University's objective of excellence in research,
scholarship, and education by publishing worldwide. Oxford is a
registered trade mark of Oxford University Press in the UK and in
certain other countries

British Library Cataloguing in Publication Data
Data available

978-0-19-841275-5

3 5 7 9 10 8 6 4 2

Paper used in the production of this book is a natural, recyclable
product made from wood grown in sustainable forests.
The manufacturing process conforms to the environmental
regulations of the country of origin.

Printed in Great Britain by Ashford Colour Press Ltd.

Acknowledgements
The publishers would like to thank the following for permissions to
use copyright material:

Cover illustrations: Istock/Iconeer

Contents

AS Level

A Level

AS Level and A Level

Introduction

The *Exam Success* series has been designed to help you reach your highest potential and achieve the best possible grade. Each book fully covers the syllabus and is written in syllabus order to help you prepare effectively for the exam. In contrast to traditional revision guides, these new books contain advice and guidance on how to improve answers, giving you a clear insight into what examiners are expecting of candidates. There is a unique section on how you can raise your achievement with advice on how to make the best use of your study time, how to avoid the most common weaknesses in examination answers and how best to prepare for the examination. Each of the *Exam Success* titles consists of two parts: the first part covers the content of the subject, while the second part contains exam-style questions and advice on how to improve your performance in the exam. All of the titles are written by authors who have a great deal of experience in knowing what is required of candidates in exams.

Exam Success in Cambridge International AS and A Level Accounting has been written specifically to meet the requirements of the AS Level and A Level Cambridge 9706 Accounting syllabus. The first part of the book consists of 20 units. The first 10 of these cover the content that is required for the AS exam, while the last 10 cover the additional content required for the A Level exam. At the start of each unit, a page reference is given to the corresponding section from the OUP student book *Accounting for Cambridge International AS & A Level*, should you wish to study the topic in more depth. Each unit closely follows the syllabus and includes illustrations, exam tips and worked examples throughout. There is a particular focus in each unit on the more important and challenging techniques. A revision checklist is provided at the end of each unit, which you can use to ensure you have covered all of the topics. There are also examples of exam-style questions at the end of each unit including a variety of structured questions. In the second part of the book there are additional units which include advice on how to make sure your study of accounting is really effective and how to ensure you perform well in the examination (Unit 21) and further examples of exam-style questions including multiple choice questions and structured questions (Unit 22).

Each of the titles in the *Exam Success* series contains common features to help you do your best in the exam. These include the following:

Key terms

These give you easy-to-understand definitions of important terms and concepts of the subject.

✗ Common errors

These give you a clear indication of some of the errors that students have made in exams in the past, helping you to avoid making the same kinds of mistakes.

★ Exam tips

These give you guidance and advice to help you understand exactly what the examiners are looking for from you in the exam.

✓ What you need to know

These provide you with useful summaries of the main features of topics you would need to demonstrate an understanding of in the exam.

💡 Remember

These include key information that you will need to remember in the exam if you are to achieve your highest possible grade.

Raise your grade

In these sections, you will be able to read answers by candidates who do not achieve maximum marks. This feature provides advice on how to improve the grade for these answers. To make the most of this feature, it is suggested that you try the exam-style question first and then compare your answer with the candidate's answer to see if you can identify any weaknesses. You can then compare your answer with the model answer to see how your own answer could have been improved. This proactive approach will help you maximise the effectiveness of this important section in each of the first 20 units and could significantly improve your skills.

Revision checklist

Each unit has a revision checklist, which gives you the opportunity to check whether you have fully understood all of the material that is covered in the unit.

Exam-style questions

Each unit contains examples of exam-style questions so you understand the type of question to expect in your exam. There is also a complete unit of exam-style questions towards the end of the book in Unit 22, which are arranged by paper. These questions provide opportunities to practise the skills and techniques required of you in the exam.

 The answers to all of the exam-style questions included in the book can be found on the OUP support website.

The *Exam Success* series is clearly focused on giving you practical advice that will help you to do your best in the exam. The model candidate answers in particular, are extremely useful as they include examiner commentary and feedback. The expectations of examiners are made very clear throughout the books, so you can fully understand what is expected of you in the exam. This will also help to boost your confidence as you approach the exam.

 Access your support website for additional content here:
www.oxfordsecondary.com/9780198412755

1 Double entry and accounting concepts

AS 1.1.1–1.1.5 SB pages 8–68

Key topics

- the accounting cycle
- source documents and the books of prime entry
- ledger accounts
- accounting concepts
- greatest challenges
- important techniques
- preparing entries in the general journal
- capital and revenue items.

✓ What you need to know

This unit reviews the key features of both the double entry model and also the accounting concepts. These features are fundamental to real success in all aspects of financial accounting because they govern how financial details should be recorded.

1.1 The accounting cycle

The key elements in the accounting cycle are illustrated below:

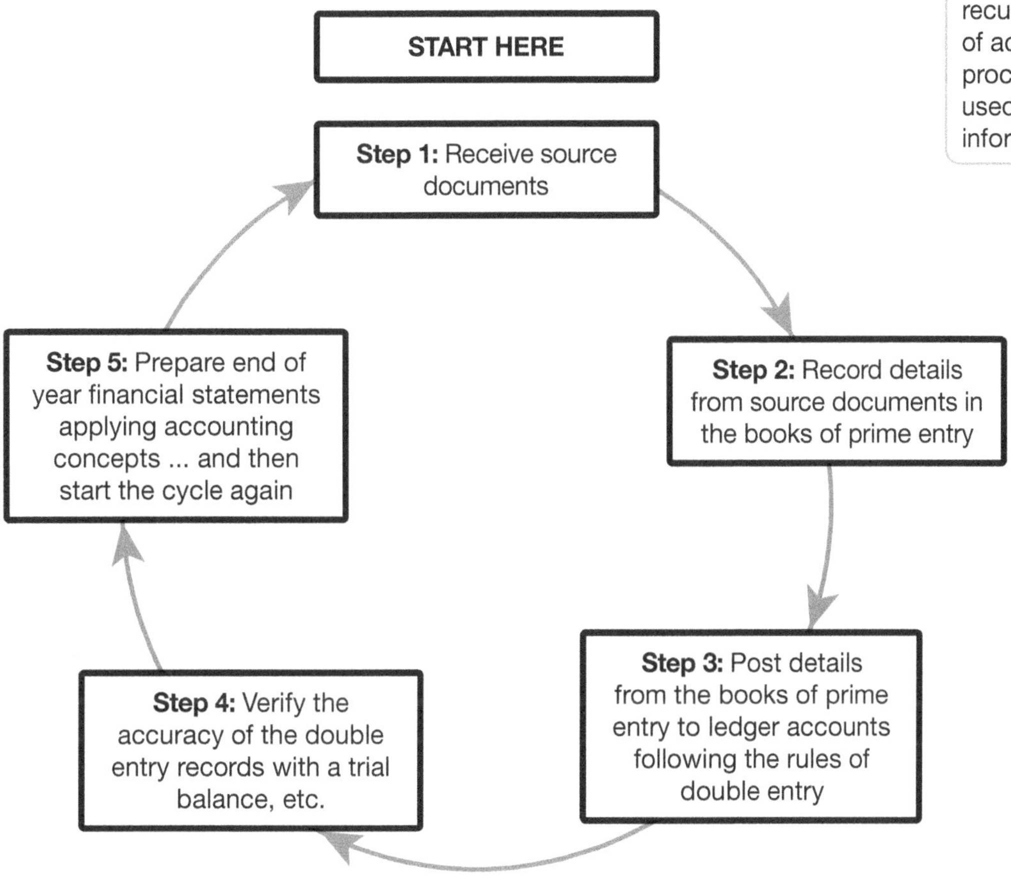
START HERE
Step 1: Receive source documents
Step 2: Record details from source documents in the books of prime entry
Step 3: Post details from the books of prime entry to ledger accounts following the rules of double entry
Step 4: Verify the accuracy of the double entry records with a trial balance, etc.
Step 5: Prepare end of year financial statements applying accounting concepts ... and then start the cycle again

▲ **Figure 1.1** The accounting cycle

Key term

Accounting cycle: the recurring sequence of activities and procedures which are used to record financial information.

AS Level

1.2 Source documents and books of prime entry

Table 1.1 lists the main source documents with links to books of prime entry.

▼ **Table 1.1** Source documents and books of prime entry

Source document	Type of transaction	Books of prime entry
Purchase invoice	Purchase of goods on credit for resale	Purchases journal
Sales invoice	For sale of goods on credit	Sales journal
Credit note received	Received from a supplier when goods are returned, i.e. returns outwards	Returns outwards journal
Credit note issued	Sent to a customer when goods are returned, i.e. returns inwards	Returns inwards journal
Paying-in slip counterfoil	Details of cash and cheques paid into a bank account	Cash book
Cheque counterfoil	Details of payments made by cheque	Cash book
Bank statement	Details of bank charges, dishonoured cheques, etc.	Cash book
Till roll	Cash sales	Cash book
Petty cash voucher	Details of small cash payments	Petty cash book
Invoice	For purchase and sale of non-current assets on credit	General journal

The three key elements of the accounting equation are: assets, liabilities and capital and the equation is:

Assets = Capital + Liabilities

The double entry model is based on the accounting equation and recognises that for any transaction two entries will be required: a debit entry and a credit entry.

▼ **Table 1.2** Guide to making double entry records

Basic transactions	*Debit entries required for:*	*Credit entries required for:*
Changes in asset values	Increase in an asset	Decrease in an asset
Changes in liability values	Decrease in liability	Increase in liability

Key terms

Source documents: supporting records that provide evidence of all the transactions to be recorded in an accounting system.

Books of prime entry: these books list the relevant information taken from source documents in date order and provide the information required to prepare the double entry records.

Remember

Record *totals* from the purchases, sales and returns journals in the general ledger accounts and not the detail of each transaction.

X Common error

Recording the totals of the discount columns in the cash book to the wrong sides of the discounts accounts in the general ledger.

Key term

The accounting equation: a formula which links the three key elements in accounting records.

X Common error

Recording the details from a source document in the wrong book of prime entry – for example, recording the invoice for a non-current asset in the purchases journal rather than the general journal.

Changes in capital (the net value of the business)	Drawings (decreasing the capital of the business)	Additions to capital
	Expense payments (reducing the net value of the business)	Sales and other incomes (increasing the net value of the business)
More technical matters		
Transferring costs and expenses to the income statement	Income statement	Expenses, purchases, etc. to transfer appropriate amount
Transferring revenue and other incomes to the income statement	Sales and other income accounts (e.g. discount received) to transfer the appropriate amount	Income statement
Create a provision for depreciation, provision for doubtful debts, etc.	Income statement	Provision accounts

1.3 Ledger accounts

Ledger accounts are usually divided into three categories:

➤ **Receivables ledger**: for the personal accounts of customers to whom goods have been sold on credit.

➤ **Payables ledger**: for the personal accounts of suppliers who have provided a business with goods on credit.

➤ **General ledger** (sometimes called the nominal ledger): for other accounts sometimes described as real account and nominal accounts.

1.4 Accounting concepts

Accounting concepts provide guidance to the bookkeeper and accountant on how to record financial information including more unusual or complex matters. Having clearly established concepts means that financial statements will be the same, no matter who prepares them.

Key terms

Real accounts: the accounts of assets.

Nominal accounts: the accounts of expense and income items.

Accounting concepts: the rules which have been developed to ensure that accounting records are prepared in a uniform way.

▼ **Example 1.1** Accounting concepts

Concept	Description	Some implications
Key concepts		
True and fair view	Financial statements should record factual information and only show reasonable estimates if necessary.	Income statements should show a true and fair profit or loss and statements of financial position should show a true and fair view of the business's financial position.
Duality	Every transaction has two aspects.	Every transaction is recorded by making both a debit entry and a matching credit entry.

Consistency	Policies adopted for recording financial details should continue to be used year-on-year.	Consistent use of the same policies ensures financial results are capable of valid comparison.
Business entity	Only matters affecting the financial position of a business should be recorded in that business's accounts.	Private transactions affecting the owner of a business are not recorded in the accounts of the business owned by that individual.
Accruals (also called matching)	When calculating profits and losses the costs and revenues for the accounting period should be matched.	Adjustments should be made for prepayments, accruals, income due, income received in advance, opening and closing inventories, etc.
Money measurement	Financial statements record only those details which have a monetary value.	Financial statements ignore some key features of a successful business: staff motivation, location, management expertise, etc.
Other concepts		
Going concern	The assumption that a business entity will continue in existence for the foreseeable future (at least one year).	Assets should be valued on the basis of their cost; the resale values of assets should be ignored as they are irrelevant.
Historic cost	The original cost of purchase should be used as the basis of valuation.	Ensures that an individual's opinion of a valuation can be ignored; valuations are based on factual evidence.
Materiality	Financial statements should not take account of items that are trivial or which could be misleading.	Financial statements record information which is significant to the users of those statements.
Prudence	Losses are accounted for as soon as they are anticipated, but profits should not be recognised until they are realised.	By anticipating losses but not recognising profits until they are realised, financial statements do not provide an over-optimistic view of a business's financial situation which could be misleading.
Realisation	Revenue should only be recorded when it has been received or when an invoice has been issued.	Sales on credit are ignored until an invoice is issued as until this point the trade receivable is under no obligation to pay.
Substance over form	The emphasis is on recording realities of a financial situation.	Sometimes the legal form of a financial situation is ignored and the real substance recorded instead.

★ Exam tip

If you are asked to explain how an accounting concept should be applied to a particular situation, include the reason for the concept in your answer as well as describing the rule.

1.5 **What are the greatest challenges?**

➤ Preparing accurate double entry records.

➤ Recording details (dates, narratives, etc.) correctly within ledger accounts and books of prime entry.

➤ Balancing and closing accounts correctly.

➤ Identifying and applying accounting concepts.

➤ Correctly distinguishing between capital expenditure and revenue expenditure, and between capital receipts and revenue receipts.

1.6 **Review some important techniques**

Recording details in ledger accounts

Every entry in a ledger account requires a date, a narrative and an amount. It is recommended that the first date on each side of a ledger account should include a year as well as the month and day. The narrative should refer to the source of the information (i.e. one of the books of prime entry) or the ledger account in which the matching entry has been made.

Balancing and closing accounts

At the end of financial period all ledger accounts should either be:

➤ balanced – where there is an amount left in the account

➤ closed – where there is no balance (often because the amount in the account has been transferred elsewhere in the accounting system).

▼ **Example 1.2** The process of balancing and closing ledger accounts

Accounts with a closing balance

Step 1: recording the balance to carry down (recorded on the side with the lower total).

Dr			Trade receivable: Yasmin Sahera				Cr
2016				2016			
May	1	Balance	860	May	10	Returns in	70
	8	Sales	440		12	Bank	620
					31	**Balance c/d**	**610**

Dr			Bank loan				Cr
2016				2016			
May	21	Bank	900	May	1	Balance	8 600
	31	**Balance c/d**	**7 700**				

Step 2: recording totals on each side of the account (totals should be on the next available *line and should be at the same horizontal level).*

Dr			Trade receivable: Yasmin Sahera				Cr
2016				2016			
May	1	Balance	860	May	10	Returns in	70
	8	Sales	440		12	Bank	620
					31	Balance c/d	610
			1 300				**1 300**

Dr			Bank loan				Cr
2016				2016			
May	21	Balance	900	May	1	Balance	8 600
	31	Balance c/d	7 700				
			8 600				**8 600**

Step 3: recording the balance brought down (the balance c/d will be a matching entry for the balance to carry down to follow the rules of double entry; the date used should be for the first day of the next month).

Dr			Trade receivable: Yasmin Sahera				Cr
2016				2016			
May	1	Balance	860	May	10	Returns in	70
	8	Sales	440		12	Bank	620
					31	Balance c/d	610
			1 300				**1 300**
June	**1**	**Balance b/d**	**610**				

Dr			Bank loan				Cr
2016				2016			
May	21	Balance	900	May	1	Balance	8 600
	31	Balance c/d	7 700				
			8 600				8 600
				June	**1**	**Balance b/d**	**7 700**

Accounts with just one entry do not have to be balanced.

Accounts with no closing balance should be closed by the use of total lines.

After closing the accounts use totals where there are several entries; use single total lines where there is just one entry on each side.

Dr			Drawings				Cr
2016				2016			
May	1	Balance	7 400	May	31	Capital	8 320
	22	Bank	920				
			8 320				8 320

Dr	Trade payable: Triumph Supplies		Cr
2016		2016	
May 24 Bank	1100	May 1 Balance	1100

> **✗ Common error**
>
> Failing to record details correctly in ledger accounts – dates, narratives, etc.
>
> Using incorrect techniques to balance accounts.

1.7 Preparing entries in the general journal

Entries in the general journal should consist of:

➤ the date

➤ the account to be debited and the amount to be debited

➤ the account to be credited and the amount to be credited

➤ a narrative explaining the reason for the entries (but note that sometimes examination questions state that narratives are not required).

Entries are made in the general journal for any transaction or event which cannot be entered in the other six books of prime entry. For example, for transactions which do not involve cash and bank accounts, and which do not involve credit transactions for purchases, sales or returns of goods in which the business trades. The following list identifies some of the main uses of the general journal:

➤ setting up a new accounting system

➤ purchases and sales of non-current assets on credit

➤ withdrawal of inventory by the owner of the business

➤ transfers of information between accounts (e.g. transferring information to the income statement) (see also Unit 4)

➤ correcting errors (see also Unit 3).

1.8 Capital and revenue items

It is important to be able to distinguish between capital and revenue items. Any misjudgement will lead to errors in preparing a business's income statement and statement of financial position. Since both these statements are used in assessing performance leading to important decisions, it follows that any errors could have serious consequences.

> **★ Exam tip**
>
> Don't forget to bring down the balance on an account – the mark for the balance is usually given for the balance brought down (not the balance to carry down).

> **💡 Remember**
>
> Provide an adequate narrative for entries in the general journal unless the question states that this is not required.

	Definition	Examples
Capital expenditure	Payments made to acquire non-current assets – i.e. payments made which will benefit the business in the long term (more than one year).	Purchase of motor vehicle. Amounts paid to upgrade/improve a non-current asset or bring into a condition where it will be useful to the business.

Revenue expenditure	Payments made for running costs – i.e. payments made which will benefit the business in the short-term (less than one year).	Day-to-day expenses such as wages, rent, utility costs, purchases of goods for resale etc.
Capital receipts	Receipts which do not arise from the normal business activities.	Capital contributions by the owner(s). Long-term loans. Proceeds from the sale of non-current assets.
Revenue receipts	Receipts arising from normal business activities or from activities incidental to normal business activities.	Income from sales. Rent receivable.

Special note

In the examinations it is not usual to have questions which focus specifically on the preparation of double entry records. It is often assumed that candidates will have mastered the techniques required in previous studies. Therefore, no example is given here of questions and of a candidate's answers to basic double entry processes. However, because double entry skills are fundamental to all financial accounting, there is a lengthy illustration demonstrating the entire accounting cycle and including some techniques from later units in this book on the online support website. There is also a model answer together with some notes which draw attention to some aspects of the topic which are often not done well.

 Access your support website for additional content here:
www.oxfordsecondary.com/9780198412755

Key topics

➤ purchasing a non-current asset

➤ depreciation

➤ depreciation calculations using the straight-line, reducing balance and revaluation methods

➤ accounting concepts in respect of non-current assets

➤ greatest challenges

➤ important techniques

➤ calculating depreciation charges.

✓ What you need to know

This unit covers transactions involving non-current assets, the need for depreciation of assets and how to calculate depreciation.

> **Key term**
>
> **Depreciation**: the loss of value of a non-current asset over its useful economic life.

2.1 Purchasing a non-current asset

The purchase of a non-current asset is an example of capital expenditure as the benefit to be gained from the purchase is intended to last for more than one accounting period. The purchase should be recorded in the cash book (if paid for immediately) or in the general journal (if the purchase is on credit). The purchase cost should include any expenses incurred necessary to ensure the asset can be put into use. For example, the cost of property should include payment for any legal costs; the cost of a machine should include the cost of delivery, plus any installation costs.

2.2 Depreciation

Non-current assets are subject to a decline in value because of:

➤ wear and tear

➤ obsolescence

➤ technological changes

➤ time factors

➤ inadequacy

The only asset which is not subject to depreciation is land, as this is the one asset with an indefinite life.

2.3 Depreciation calculations

There are three accepted methods of calculating depreciation:

➤ Straight-line method: in many cases the annual depreciation charge is expressed as a percentage of cost. So a depreciation charge of 20 per cent per annum would be applied if the asset's useful life was five years. This method has the advantage of simplicity as it makes it relatively easy to calculate the effect of depreciation charges on future profit projections. However, this method overlooks the fact that many assets lose value more quickly in the first years.

> **Key term**
>
> **Straight-line method**: the cost of the asset less the expected residual value, spread equally over the estimate useful life of the asset.

> **Reducing balance method**: this method involves more complicated calculations, but it recognises the fact that some assets (particularly machinery and vehicles) lose more value in the first years. When increasing repair costs are taken into account, the reducing balance method can lead to a more even annual charge to set against revenue.

> **Revaluation method**: is used where the asset consists of a collection of many items each of which has cost relatively little (e.g. tools in an engineering company; crockery in a hotel).

Depending on the business's depreciation policy, it may be necessary to calculate depreciation on a non-current asset which was newly acquired part-way through a year, it may also be necessary to calculate depreciation for part of a year on a non-current asset sold part-way through a year.

2.4 Accounting concepts and non-current assets

> **Cost concept**: applies to the value of each non-current asset to ensure that non-current assets are valued objectively.

> **Going concern concept**: confirms that non-current assets should be valued on the basis of their cost, rather than their possible resale value, since it is assumed that the business will continue to trade and so the assets will continue to be used by the business.

> **Accruals (matching) concept**: applies to charging depreciation each financial year to ensure profit and loss calculations take account of the resources used during the year to generate revenue for the business.

> **Consistency concept**: applies to the application of the method of depreciation. Once a particular method and rate of depreciation has been adopted (for example, straight-line) it continues to be used, to ensure that valid comparisons of financial statements can be made year on year. Of course, depreciation policies can be changed if circumstances change, but the users of the financial statements need to be fully informed of the impact of the change. (They can then make adjustments in order to account for the change.)

Disposal of non-current assets

The disposal of a non-current asset requires the preparation of an asset disposal account which will show entries for the following:

> transfer of the asset's original cost

> transfer of the asset's accumulated depreciation

> the amount received at the time of the disposal

> the profit or loss on the disposal.

Complications can arise because it may be necessary to calculate depreciation on the asset being sold up to the time of the disposal (this will depend on the business's depreciation policy), and the disposal could involve a part exchange arrangement.

> **Key terms**
>
> **Reducing balance method**: a fixed percentage of the net book value (carrying amount) of the asset is charged as depreciation each year.
>
> **Revaluation method**: the annual depreciation charged is based on comparing the estimated value of the asset at the beginning and at the end of the year.

2.5 What are the greatest challenges?

➤ Making accurate records in ledger accounts of non-current assets and depreciation.

➤ Preparing correctly detailed journal entries relating to non-current assets.

➤ Correctly calculating depreciation charges.

➤ Recording an asset disposal, particularly where there is a part-exchange arrangement.

2.6 Review some important techniques

Journal and ledger records for non-current assets

There will be separate accounts to record the cost of each type of non-current asset and separate accounts to record the provision for depreciation on each type of non-current asset. As with all ledger accounts, it is important to record the dates and narratives correctly.

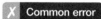 **Common error**

Presenting information in the wrong order in a journal entry – the debit entry should be first and the credit entry second.

▼ **Example 2.1** Ledger records for non-current assets

Rahul purchased a new machine for business use on credit from QTQ Engineering on 1 April 2015. The machine cost $34 200. In addition, Rahul paid the supplier $1300 to have the machine delivered and installed. Rahul's policy is to depreciate non-current assets by 20 per cent per annum. The business's accounting year ends on 31 March.

The journal entries to record the purchase of the machine and the first year's depreciation are as follows:

GENERAL JOURNAL

				Dr $	Cr $
2015					
April	1	Machinery		35 500	
		QTQ Engineering			35 500
		purchase of machine including installation costs on credit			
2016					
March	31	Income statement		7 100	
		Provision for depreciation (machinery)			7 100
		annual provision of depreciation on machinery			

The ledger records are shown below:

Dr	Machinery account		Cr
2015			
April 1 QTQ Engineering	35 500		

Dr	Provision for depreciation (Machinery) account		Cr
	2016		
	March 31 Income statement	7 100	

Calculating depreciation charges

Normally, calculating an annual depreciation charge is straightforward using the three depreciation methods. However, complications can arise when a business has purchased or sold an asset during an accounting period and has a policy of providing depreciation on a month-by-month basis.

> **▼ Example 2.2** Calculating depreciation charges
>
> An engineering business was established on 1 January 2015 with the following non-current assets:
>
> | | $ |
> |---|---|
> | Plant and machinery | 108 000 |
> | Equipment | 45 000 |
> | Loose tools | 9 500 |
>
> During the year additional loose tools were purchased for $2900.
>
> At 31 December depreciation is to be provided as follows:
>
> ➤ Plant and machinery: at 20 per cent per annum using the straight-line method, taking account of an estimated residual value of $4500.
>
> ➤ Equipment: at 30 per cent per annum using the reducing balance method.
>
> ➤ Loose tools to be revalued at $10 800.
>
> Depreciation charges for the year ended 31 December 2015 are as follows:
>
> | | Charge $ | Calculation details |
> |---|---|---|
> | Plant and machinery | 20 700 | 20% × (cost less residual value) i.e. 20% × ($108 000 – $4 500) |
> | Equipment | 13 500 | 30% × cost $45 000 |
> | Loose tools | 1 600 | Original cost $9 500 + additions $2900 – closing valuation of $10 800 |
>
> During 2016 additions were:
>
> ➤ Loose tools $4100;
>
> ➤ Plant and machinery: $8500 (residual value $900) purchased on 1 October 2016.
>
> Depreciation of plant and machinery is made on a month-by-month basis.
>
> Loose tools were value at $11 300 on 31 December 2016.
>
> Depreciation charges for the year ended 31 December 2016 are as follows:
>
> | | Charge $ | Calculation details |
> |---|---|---|
> | Plant and machinery | 21 080 | Original asset: $20 700

Addition: 20% × (cost less residual value, i.e. $7600 × ¼ (i.e. 3 months) = $380 |
> | Equipment | 9 450 | 30% × nbv on 1 January 2016 (cost $45 000 – provision for depreciation $13 500) |
> | Loose tools | 3 600 | Value 1 January 2016 $10 800 + additions $4 100 – closing valuation of $11 300 |

Remember

Check the business's depreciation policy to see if depreciation should be calculated in the year of purchase or year of sale of a non-current asset.

▼ **Example 2.3** Preparing a provision for depreciation account over a number of years

Using the details about plant and machinery in Example 2.2 on the previous page, the provision for depreciation account will appear as follows:

Dr			Provision for depreciation (Plant and machinery) account				Cr
2016				2015			
Dec	31	Balance c/d	41 780	Dec	31	Income statement	20 700
				2016			
				Dec	31	Income statement	21 080
			41 780				41 780
				2017			
				Jan	1	Balance b/d	41 780

▼ **Example 2.4** Disposal of a non-current asset on a part exchange basis

Majid purchased a new motor vehicle on 1 August 2016 with a total cost of $30 000. He paid by cheque, $17 500 and part exchanged his old vehicle for the balance. The old vehicle had cost $24 000 when purchased on 1 March 2014. Majid's policy is to depreciate all non-current assets at 20 per cent per annum using the straight-line method on a month-by-month basis, but no depreciation is provided on an asset in the year of sale.

Here are the ledger accounts to record this information covering each of the years 2014, 2015 and 2016.

Dr			Motor vehicles account				Cr
2014				2016			
March	1	Bank	24 000	Aug	1	Motor vehicle disposal	24 000
2016				2016			
Aug	1	Disposal	12 500	Dec	31	Balance c/d	30 000
	1	Bank	17 500				
			30 000				30 000
2017							
Jan	1	Balance b/d	30 000				

Dr			Provision for depreciation (Motor vehicles) account				Cr
2015				2014			
Dec	31	Balance c/d	8 800	Dec	31	Income statement	4 000
				2015			
				Dec	31	Income statement	4 800
			8 800				8 800
2016				2016			
Aug	1	Motor vehicle disposal	8 800	Jan	1	Balance b/d	8 800
Dec	31	Balance c/d	2 500	Dec	31	Income statement	2 500
			11 300				11 300
				2017			
				Jan	1	Balance b/d	2 500

Dr		Motor vehicles disposal account				Cr
2016				2016		
					Motor vehicles: provision for	
Aug	1 Motor vehicles	24 000	Aug	1	depreciation	8 800
				1	Motor vehicles	12 500
				1	Income statement	2 700
		24 000				24 000

Remember

In order to record the part-exchange element it is necessary to make the following entries in the motor vehicles (at cost) account.

➤ Debit the account with the amount paid for the new vehicle (and credit the bank account).

➤ Debit the account with a balancing amount to ensure the total value of the new vehicle emerges as the balance on the account (and credit the disposal account with this amount). In the example the new vehicle is worth $30 000 so this must emerge as the balance on the account. The amount paid is $17 500. So the transfer to the disposal account is $12 500.

Revision checklist

I can:

➤ define and explain key terms ☐

➤ state the causes of deprecation ☐

➤ explain how accounting concepts are applied to non-current assets ☐

➤ prepare journal entries to record transactions concerning non-current assets ☐

➤ explain depreciation methods and calculate depreciation charges ☐

➤ prepare correctly detailed ledger accounts for non-current assets and depreciation ☐

➤ record disposals of non-current assets including situations involving a part exchange. ☐

⬆ Raise your grade

Iqbal purchased a delivery vehicle for his business on 1 April 2014. The vehicle cost $24 800. Iqbal paid $1 600 to have shelving fitting in the vehicle. He estimates the vehicle will have a useful life of four years and a residual value of $2 400. On 1 October 2016 he part exchanged the vehicle for a larger delivery vehicle with a cost of $36 000, paying $24 000. The new vehicle has estimated residual value of $5 400 and a useful life of five years. Iqbal depreciates non-current assets on a month-by-month basis, but does not charge depreciation in the year of disposal. Payments were made by cheque. The business's accounting year ends on 31 December.

Prepare entries in general ledger accounts necessary to record the purchase, depreciation and eventual disposal of a non-current asset involving a part exchange arrangement. The accounting records are required to cover the years 2014, 2015 and 2016.

Student answer

Dr			Delivery vehicle account				Cr	
	April	1	Bank	24 800	Oct	1 Disposal	26 400	
❶		1	Bank	1 600				❷
❸ ❻	Oct	1	Bank	36 000				

Dr		Provision for depreciation (delivery vehicle) account				Cr			
Oct	1 Disposal	12 000	Dec	31	Depreciation	6 000	❹	❺	❼
❷				31	Depreciation	6 000	❶		
							❺		

Dr			Disposal (delivery vehicle) account				Cr	
Oct	1	Delivery vehicle	26 400	Oct	1 Prov deprcn	12 000		
					Income			
				1 stmnt	14 400	❻	
			26 400			26 400		

How to improve this answer

❶ Dates should include years; correct narratives are not always shown.

❷ All the accounts should be balanced and the balances brought down.

❸ The rules of double entry have not always been correctly applied.

❹ Depreciation charge for the first year has not been correctly calculated.

❺ Depreciation policies not correctly applied.

❻ Part exchange transaction has not been correctly recorded.

❼ Absence of workings for the more complex calculations.

Model answer

Dr			Delivery vehicle account				Cr
2014				2014			
April	1	Bank	24 800	Dec	31	Balance c/d	26 400
	1	Bank	1 600				
			26 400				26 400
2015				2016			
Jan	1	Balance b/d	26 400	Oct	1	Disposal	26 400
				Dec	31	Balance c/d	36 000
2016							
Oct	1	Disposal	12 000				
	1	Bank	24 000				
			62 400				62 400
2017							
Jan	1	Balance b/d	36 000				

Dr			Provision for depreciation (delivery vehicle) account				Cr
2015				2014			
Dec	31	Balance c/d	10 500	Dec	31	Income st (W1)	4 500
				2015			
				Dec	31	Income st (W2)	6 000
			10 500				10 500
2016				2016			
Oct	1	Disposal	10 500	Jan	1	Balance b/d	10 500
Dec	31	Balance c/d	1 530	Dec	31	Income st (W3)	1 530
			12 030				12 030
				2017			
				Jan	1	Balance b/d	1 530

W1: Depreciation = [($26 400 – 2 400) × 25%] × = $4 500

W2: Depreciation = ($26 400 – 2 400) × 25% = $6 000

W3: Depreciation = [($36 000 – 5 400) × 20%] × = $1 530

Dr			Disposal (delivery vehicle) account				Cr
2016				2016			
Oct	1	Delivery veh	26 400	Oct	1	Prov deprcn	10 500
					1	M Veh (part exch)	12 000
					1	Income statement	3 900
			26 400				26 400

? Exam-style questions

1   State what is meant by 'useful economic life'.

2   Explain how accounting concepts are applied when depreciating non-current assets.

3   Advise the owner of a business whether to use the straight-line or reducing balance method to depreciate a motor vehicle.

4   Amina owns a hotel. The following information is available concerning the business's non-current assets at 1 January 2016.

The motor vehicle was sold in part exchange for a new vehicle cost $38 500 on 1 May 2016. The amount paid was $20 000.

Crockery was valued at $6 550 on 31 December 2016.

(a) Calculate the depreciation charges for the year ended 31 December 2016 for:

● Freehold buildings.

● Equipment.

● Crockery.

Non-current asset	Cost or value	Date of purchase or valuation	Depreciation method and rate
Freehold buildings	$800 000	Purchased 1 January 2012	Straight-line method 2% per annum
Motor vehicle	$36 000	Purchased 1 August 2014	Straight-line method 15% per annum
Equipment	$25 000	Purchased 1 January 2015	Reducing balance method 20% per annum
Crockery	$6 200	Value at 1 January 2016	Revaluation method

Amina's policy is to depreciate non-current assets for each month of ownership except in the year of disposal.

During the year ended 31 December 2016 additional crockery was purchased at a cost of $1220.

(b) Prepare the following ledger accounts covering the years 2014, 2015 and 2016:

● Motor vehicles (at cost) account.

● Provision for depreciation (motor vehicles) account.

● Disposal (motor vehicles) account.

Key topics

➤ trial balance

➤ bank reconciliation statement

➤ control accounts

➤ reconciliation, verification and accounting concepts

➤ greatest challenges

➤ important techniques.

✓ What you need to know

How to prepare trial balances, correct errors, make use of suspense accounts, reconcile cash book and bank statement balances, prepare control accounts.

3.1 Trial balance

The trial balance is used to check the arithmetical accuracy of the double entry records; in addition it provides a useful summary of all the balances in the accounting system at a particular date which can then be used to assist in the production of financial statements. It is important to remember that agreement of trial balance totals does not necessarily guarantee the accuracy of the double entry records (see error correction below) and this is a major limitation of this technique. Where the totals of a trial balance do not match, an additional entry is made to equalise the totals and is labelled suspense account.

Account balances are shown in a trial balance according to the following model:

Key term
Trial balance: a statement listing all the balances in the accounts of an organisation on a particular date.

▼ **Table 3.1** Trial balance

Debit side	Credit side
Balances of the following accounts	*Balances of the following accounts*
Assets	Liabilities
Expenses	Revenue (sales)
Drawings	Other income (e.g. discounts received)
Purchases	Provisions
Returns inwards	Returns outwards
Suspense account (where credit total of trial balance exceeds debit total)	Suspense account (where debit total of trial balance exceeds credit total)
Current account (when negative balance – partnership)	Capital account (sole trader)
	Capital accounts (partnership)
	Current account (when positive balance – partnership)
	Share capital account (limited company)
	All reserve accounts (limited company)

3.2 Bank reconciliation statement

When applying the bank reconciliation process, the cash book is first updated to include any items shown on the bank statement which may have been overlooked. Where the two balances do not agree there will be errors in the cash book and/or bank statement records which will require further investigation. The great advantage of this technique is that a business's records (of bank transactions) can be checked with a totally independent source (the bank's records). As a result, not only is it possible to detect the presence of errors but also to reduce the chance of fraud.

Updating the cash book requires the inclusion of any item shown in the bank statement which has yet to appear in the cash book:

➤ Debit the bank columns in a business's cash book with credit transfers not yet recorded (e.g. credit customers settling amounts due by transferring funds directly to the business's bank account).

➤ Credit the bank columns with payments not yet recorded for bank charges, direct debit and standing order payments overlooked.

The reconciliation statement will take account of:

➤ Unpresented cheques (payments made by cheque and shown in the cash book which have not yet been presented for payment).

➤ Outstanding bankings (cash and cheques received by a business and included in the cash book but which have not yet been recorded by the bank).

The statement can be set out in one of two ways.

> **Key term**
>
> **Bank reconciliation**: a process for ensuring that a business's cash book bank balance can agree with that shown in a bank statement on a particular date.

> ☒ **Common error**
>
> Confusion over which items affect the cash book and which items affect the bank reconciliation statement.

> 💡 **Remember**
>
> Take care to get the arithmetical direction correct in a reconciliation statement when you start with an overdrawn balance.

Starting with updated cash book balance:

Bank Reconciliation Statement at (date)	
	$
Balance as per cash book	xxx
Add: unpresented cheques	xxx
	xxx
Less: outstanding bankings	(xxx)
Balance as per bank statement	xxx

Starting with the bank statement balance:

Bank Reconciliation Statement at (date)	
	$
Balance as bank statement	xxx
Add: outstanding bankings	xxx
	xxx
Less: unpresented cheques	(xxx)
Balance as per cash book	xxx

3.3 Preparing control accounts

> **Key term**
>
> **Control accounts**: a process which uses totals from books of prime entry to provide a summarised version of the information recorded in the sales ledger and purchases ledger. This can then be used to check the arithmetical accuracy of these ledgers.

A sales ledger account resembles a total trade receivables account and entries are made for:

▼ Table 3.2 Sales ledger account

Item	Explanation	Source of information
Debit side		
Opening balance	Total amounts owed by trade receivables at beginning of period.	Balance brought down from previous month's control account.
Credit sales	Total credit sales for the period.	Sales journal
Refunds	Total amounts refunded to credit customers who have overpaid.	Cash book
Returned cheques	Total for the period of cheques received from trade receivables but returned by the bank.	Cash book
Interest charges	Total of interest charged to credit customers on overdue accounts for the period.	General journal
Credit side		
Opening balance	Total of any credit balances in the sales ledger arising from customer overpayments.	Balance brought down from previous period's control account.
Returns inwards	Total returns inwards for the period.	Returns inwards journal
Bank	Total receipts from trade receivables for the period.	Cash book
Discounts allowed	Total of discounts allowed to trade receivables during the period.	Cash book
Contras	Total for the period of any amounts set off against the accounts of trade payables who are also credit customers.	General journal
Irrecoverable debts	Total of all trade receivable accounts written off as irrecoverable during the period.	General journal

A purchases ledger control account resembles a total trade payables account and entries are made for:

Table 3.3 Purchases ledger account

Item	Explanation	Source of information
Debit side		
Opening balance	Total of any debit balances in the purchases ledger arising from overpaying credit suppliers.	Balance brought down from previous month's control account.

Returns outwards	Total of returns outwards for the period.	Returns outwards journal
Bank	Total of payments made to credit suppliers during the period.	Cash book
Discounts received	Total of discounts received from trade payables during the period.	Cash book
Contras	Total of amounts set off against trade receivable accounts who are also suppliers.	General journal
Credit side		
Opening balance	Total of amounts owing to trade payables at beginning of period.	Balance brought down from previous period's control account.
Credit purchases	Total of credit purchase for the period.	Purchases journal
Refunds	Total amounts refunded by credit suppliers whose accounts have been overpaid.	Cash book
Cancelled cheques	Total of cheques paid to trade payables not accepted by the bank or withdrawn.	Cash book

> **Remember**
>
> Make sure you are clear about which items should be entered in a sales ledger control account and which items should be entered in a purchases ledger account, it is not unusual for a mix up to occur. Don't forget that some items should not appear in either: for example cash purchases, cash sales.

Control accounts help identify that errors have occurred when the balance on a control account does not agree with the total of the personal accounts in the ledger. They provide a useful source of information for financial statements (i.e. total trade payables and total trade receivables). Control accounts and the ledgers containing personal accounts are usually prepared by different members of staff; as a result, control accounts can help in the prevention of fraud. However, they only check the arithmetical accuracy of the ledgers; they do not reveal the presence of all errors, for example where a transaction has been posted to the wrong personal account within a ledger.

3.4 Reconciliation, verification and accounting concepts

The techniques for verifying accounting records check the arithmetical accuracy of the accounting records. This means, in effect, that they check whether the duality concept has been properly applied, i.e. whether for each debit entry there is a matching credit entry. Correcting errors also requires the application of the duality concept, and requires a sound understanding of the rules of double entry.

3.5 What are the greatest challenges?

Error correction

➤ Correctly distinguishing between errors that are revealed by a trial balance and those that are not.

➤ Deciding which account should be debited and which credited when correcting an error.

- When correcting a draft profit, working out whether an error affects profit and if so, whether to increase or decrease the draft figure.

- When revising an incorrect statement of financial position, deciding which errors affect the draft profit and which do not.

Bank reconciliation

- Correctly identifying items omitted from the cash book and items yet to be recorded on the bank statement.

- Reconciling the cash book bank balance and bank statement balance when there is an overdraft.

Control accounts

- Identifying items for inclusion in a control account and those which should be excluded.

- Making entries on the correct sides in a control account.

- Correcting errors affecting a control account and the total of personal accounts from the relevant ledger.

3.6 Review some important techniques

Error correction

Error correction can involve preparing journal entries and making entries in a suspense account. It is important to understand when a suspense account is involved and when it is not. Errors which do not require an entry in a suspense account arise when there has been a matching debit and credit entry involving the wrong amount, wrong account, etc.

Here are six named errors:

- **Commission** – an entry has been made in the wrong account within the same class of accounts (e.g. debit entry for a sale on credit in a trade receivable account T Bradford instead of the account of T Bradfield).

- **Compensating** – by coincidence, an error in the amount recorded as a debit entry in one account is exactly matched by an error for the same amount as a credit entry in another account.

- **Complete reversal** – the correct amount is recorded but the account to be debited is credited, and the account to be credited is debited.

- **Omission** – a transaction is completely omitted from the books of account.

- **Original entry** – entries are made in the correct accounts, but the entries are for the wrong amount.

- **Principle** – where an entry is made to the incorrect class of account (e.g. debiting the machinery account with the cost of machinery repairs).

Errors which are revealed by a trial balance and require an entry in the suspense account are sometimes described as being 'one-sided' in nature, i.e. one aspect of the double entry is correct, but the matching entry is incorrect:

- Transposition error: in the case of either the debit or credit entries, figures are transposed (for example a transaction is entered as Debit $4356 and Credit $4365).

- Addition error: there is a miscalculation totalling a book of original entry, balancing an account, totalling a column in the trial balance.

- Posting error: one of the entries is posted to the wrong side of an account.

➤ Unequal posting error: where the amount of the debit entry does not match the amount of the credit entry.

➤ Partial omission: either the debit or credit entry is not recorded.

▼ **Example 3.1** Correcting errors

Errors	Guidance note	Correction
An invoice from a supplier, LZO Ltd, for $697 was entered in the purchases journal as $679	This is an example of an error of original entry. The trial balance will not reveal this error as both the supplier's account and the purchases account will be equally incorrect (by $18). The correction will not involve an entry in the suspense account.	Dr Purchases $18 Cr LZO Ltd $18
A credit note for $332 sent to a credit customer, Azzam Ltd, was correctly entered in the returns outwards journal, but was debited to the supplier's account as $323.	This is an example of an unequal posting, as the returns outwards account will be correctly credited with $332 (because the entry in the returns outwards journal is correct). However, the entry in the supplier's account is for the wrong amount. The trial balance totals would not agree and the error will affect the suspense account.	Dr Azzam Ltd $9 Cr Suspense $9
A payment by cheque of $2 400 for legal fees relating to the purchase of the business premises was debited to the administration expenses account.	This is an example of an error of principle. There is a matching debit and credit entry, so the trial balance totals will agree. However, the legal fees should be regarded as capital expenditure and debited to the premises account. Errors of principle always result in an entry being made in the wrong class of account (in this case an expense account rather than an asset account).	Dr Business premises $2 400 Cr Administration expenses $2 400
The receipt of a cheque from a trade receivable, AQY Ltd, for $560 was not posted from the cash book.	This is an example of an error of partial omission. The trial balance totals will not agree as there is no entry in the trade receivables account to match the entry in the cash book. The correction will, therefore, involve the suspense account and recording the missing (credit) entry in the account of the trade receivable.	Dr Suspense $560 Cr AQY Ltd $560
The owner withdrew some inventory, cost $150, for private use. The entries made were Dr Inventory $150, Cr Drawings $150.	There are several errors here, but the first point to notice is that there is a matching debit and credit entry, so the trial balance totals will agree and the suspense account will not be involved in the correction. The entry in the drawings account should be on the debit side; the credit entry should have been in the purchases account and not in the inventory account. To correct the error in the drawings account it will be necessary to cancel the wrong entry by debiting the account with $150 and then debit the account again with $150 to make the correct entry. The entry in the inventory account will be need to be cancelled (Cr inventory account) and a correct entry made in the purchases account (Cr purchases).	Dr Drawings $300 Cr Inventory $150 Cr Purchases $150

Using the information in Example 3.1, it is possible to construct a suspense account to determine the original difference in a trial balance. (It is assumed that all the errors in the books of account have been found and the year end is 31 December 2016.)

Dr			Suspense Account				Cr
2016				2016			
						Balance (difference	
Dec	31	AQY Ltd	560	Dec	31	in TB totals)	551
					31	Azzam Ltd	9
			560				560

By making entries for the corrections, the balance of the suspense account reveals the original difference in the trial balance totals. In the example the total of the debit side of the trial balance must have exceeded the total of the credit side by $551.

Correcting profits

Errors in the books of account can lead to the incorrect information being shown in the draft income statement. In these circumstances it will be necessary to prepare a statement to amend the draft profit or loss. The following is a guide to the amendments to be made.

Increasing a draft profit (or decreasing a draft loss)	Where income has been omitted or understated, or where costs have been overstated.
Decreasing a draft profit (or increasing a draft profit)	Where income has been overstated, or where costs have been omitted or understated.
Doubling the amount involved	Where an entry has been made on the 'wrong side' of the income statement, it will be necessary to double the amount involved to correct the mistake.
No effect	If the error has occurred in an item which does not appear in an income statement, no correction to the draft profit will be required.

Correcting a statement of financial position

Errors in the books of account can also affect the statement of financial position. Where the totals of a statement of financial position do not agree it is usual to equalise the totals by insertion of a suspense account either in the asset list or in the capital and liabilities list as appropriate.

Reconciling control accounts and ledgers

Where the balance of a control account does not match the total of the personal account balances in the relevant ledger, it will be necessary to attempt to reconcile the two figures by locating and correcting errors.

Errors could be of various types:

Type of error	Example	Guidance
Affecting the control account only.	The total of discounts received $486 was misread as $468.	Only the purchases ledger control account is affected by this error. The personal accounts of suppliers are affected by individual entries in the discounts received column in the cash book, not by the total.
Affecting a personal ledger account only.	A return inwards of $131 from the returns inwards journal was posted as $311.	The entry in the journal is correct, so the total of the journal which is used in the control account will be correct. The only error will occur in the account of the trade receivable.
Affecting both the control account and a personal ledger account.	An invoice for $480 received from a supplier was overlooked.	This error of omission will result in an understated total in the purchases journal used in compiling the purchases ledger control account; it will also result in an error in the account of the trade payable in the purchases ledger.
Having no effect.	Cash sales of $870 was recorded as $780 in the cash book.	Cash sales is irrelevant when totalling the balances of trade receivables and when compiling the sales ledger control account.

Revision checklist

I can:

➤ prepare a trial balance recording accurately account balances in the debit or credit column ☐

➤ compare a cash book and a bank statement to provide details to be used in the reconciliation process ☐

➤ prepare an updated cash book and a bank reconciliation statement ☐

➤ prepare control accounts including making entries for the less common items ☐

➤ reconcile control accounts and ledgers and know whether an item affects one or the other, both or neither ☐

➤ identify errors revealed by a trial balance and those which are not ☐

➤ prepare a suspense account and identify the original difference in the trial balance totals ☐

➤ prepare a statement to correct a draft profit ☐

➤ prepare a corrected statement of financial position. ☐

Raise your grade

Shakira has discovered several errors that had been made when preparing her business's accounting records for the year ended 31 December 2016. The draft profit for the year ended on this date was $27 450.

1 A sales invoice for $3600 had been wrongly entered in the books as $360.

2 The closing inventory included some items which had cost $1200. These items had become damaged and had a net realisable value of $760.

3 A purchases credit note for $220 had been completely omitted from the books.

4 Carriage inwards of $130 had been treated as carriage outwards.

5 Rent received of $1450 had been wrongly debited to rent expense account.

6 Depreciation of $2400 had been charged on equipment using the straight-line method and a rate of 20 per cent per annum. A rate of 25 per cent should have been used.

7 The provision for doubtful debts should have been decreased by $180. However, the income statement included an increase in the provision for this amount.

Prepare a statement showing an amended profit.

Student answer

Correction of profit ❶

	$
❶	27 450
Understated sales	3 240
Change in inventory valuation ❷	(760)
Understated returns out ❷	(220) ❶
Carriage expenses ❸	(130)
Rent received ❷	1 450
Depreciation charge ❹	(600)
Decrease in provision ❷	(360)
❶	30 070

How to improve this answer

❶ Presentation weaknesses:

- the title is inadequate
- the draft profit should be labelled
- the overall layout is satisfactory but could be set out more effectively
- the final result should be labelled.

❷ There are errors in correcting some of the mistakes:

- incorrect figure for inventory valuation – it is the decrease in the value of inventory which affects the draft profit
- if returns out needs to be increased, net purchases would decrease and profits would increase

- as the amount of rent received has been entered on the wrong side of the income statement, the amount needs to be doubled to correct the mistake.

- a decrease in the provision should increase profits.

③ The carriage error should not be included in the answer. The carriage has been included in the income statement and reduced profits – it was merely recorded in the wrong section.

④ The depreciation charge has been correctly recorded, but it would have been helpful to show the workings for this amount.

Model answer

Statement of corrected profit for the year ended 31 December 2016

	+	–	Total
	$	$	$
Original profit for the year			27 450
1 Understated sales	3 240		30 690
2 Revaluation of inventory		440	30 250
3 Understated returns outwards	220		30 470
4 No effect			30 470
5 Rent received wrongly recorded	2 900		33 370
6 Additional depreciation charge (W1)		600	32 770
7 Decrease in provision for doubtful debts	360		33 130
Corrected profit for the year			33 130

W1: Additional depreciation: extra 5% required; since 20% = $2400, 5% = $600

★ Exam tips

Key features of the model answer

➤ Take care to provide a full heading and label each item in the table.

➤ The best layout has separate columns to record increases, decreases and an updated total.

➤ Take care to work out the effect of any error on profits – double check that your thinking is correct.

➤ Remember that it is necessary to double the amount when correcting any error where an entry has been made 'on the wrong side'.

➤ Provide workings for any more complex calculations.

1 State what is meant by an error of original entry.

2 Describe two advantages of preparing a trial balance.

3 Explain the purpose of a suspense account.

4 The owner of business thinks that preparing bank reconciliation statements and control accounts is a waste of time and should be discontinued since errors continue to occur in the books of account. Do you agree with this view? Advise the owner as to whether the verification techniques should be continued. Justify your decision.

5 The totals of a business's trial balance did not agree and the difference was recorded in a suspense account. Subsequently the following errors were discovered in the books of account.

- The balance of the purchases account had been understated by $240.

- A credit note for $142 received from a credit customer, Xi, has been entered as $124 in the books of account.

- The total of the discounts allowed column in the cash book $475 had been credited to the discounts received account.

- A payment of $1200 for equipment repairs had been debited to the equipment account. (Depreciation of equipment can be ignored as no depreciation charges had yet been made for the year under review.)

- The owner of the business had taken goods costing $515 for private use. The only entry made for this transaction had been to credit the drawings account with $515.

Prepare entries in the general journal to correct these errors. (Narratives are not required.)

Prepare the suspense account to show the original difference in the trial balance totals.

6 A business's cash book showed an overdrawn balance of $3712 on 30 September 2016. On the same date the balance shown in the business's bank statement was $2 943 overdrawn. When the cash book and bank statement were compared the following discrepancies were discovered:

- Bank charges of $180 appeared in the bank statement but had not been entered in the cash book.

- Cheques totalling $3486 drawn by the business were unpresented.

- A direct debit for utility charges of $850 appeared in the bank statement but had been omitted from the cashbook.

- Cash sales totalling $1332 had been paid into the bank and recorded in the business's books of account, but had not yet been credited by the bank.

- The business had paid a supplier, New Supplies Ltd, $1758 by cheque. This had been recorded correctly on the bank statement, but appeared in the cash book as $1785.

- The owner of the business was not aware until the bank statement was received that the bank had returned a cheque for $382 received from a credit customer, OZQ Ltd.

Prepare an updated cash book for the month of September 2016.

Prepare a bank reconciliation statement at 30 September 2016.

7 (a) On 31 October 2016 a bookkeeper prepared a sales ledger control account from the following information:

	$
Total of balances in the sales ledger on 1 October 2016	
Debit	8 495
Credit	240
Totals from books of prime entry for October 2016	
Sales journal	32 440
Returns inwards journal	1 172
Cash book	
Receipts from credit customers	28 560
Refunds to credit customers	450
Discounts allowed	1 010
Dishonoured cheques	424
General journal	
Irrecoverable debts written off	380
Contras: purchases ledger accounts set off against sales ledger accounts	533
Interest charged on overdue balances	157
Cancellation of discount allowed	62

Prepare the sales ledger account for October 2016.

(b) On 31 October 2016 the total of balances in the sales ledger was $9143. The following errors have been discovered.

- The total of the sales journal had been overcast by $730.
- Returns inwards from credit customer ADT Ltd, had been credited to the account of AAD Ltd.
- A refund to a credit customer of $130 had been correctly recorded in the cash book, but had been credited to the account of the customer.
- A sales credit note for $145 had been entered in the returns inwards journal as $154.

Prepare statements at 31 October 2016 to show:

(i) corrected sales ledger control account balanced;

(ii) corrected total of the sales ledger accounts.

8 The following draft statement of financial position has been prepared by an inexperienced bookkeeper for the business owned by John Bapniah, a sole trader, at 30 September 2016.

Statement of financial position for y/e 31 December 2016

	$	$
Non-current assets		
Machinery and plant at net book value	78 400	
Furniture and fittings at net book value	14 900	
		93 300
Current assets		
Inventory	12 200	
Trade receivables (net of provision for doubtful debts $550)	13 450	
Other receivables	1 130	
Cash at bank	920	
		27 700
		121 000

Capital		
Balance 1 January 2016	111 100	
Draft profit for year	9 960	
Drawings	(26 240)	
		94 820
Non-current liabilities		
Loan (8%)		12 000
Current liabilities		
Trade payables	13 810	
Other payables	890	
		14 700
		121 520

The following errors were discovered:

- Included in the inventory were some damaged items which had cost $1320, but which now have an estimated sale value of $1400 after being repaired at a cost of $220.
- A credit note received from a trade supplier for $390 had been entered twice in the books of account.
- A decision to increase the provision for doubtful debts to 5 per cent of trade receivables had not been implemented.
- Rent had been correctly adjusted for a prepayment of $260. However, the prepayment had been included in other payables in the statement of financial position.
- Machinery repairs of $1450 had been debited to the machinery account in error. As a result the annual depreciation charge had been overstated by $120.
- No entries had been made for the repayment of loan, $1400, made on 31 December 2016.

Prepare a corrected statement of financial position at 31 December 2016.

4 Preparation of financial statements for a sole trader

Key topics

- ➤ preparing an income statement
- ➤ preparing a statement of financial position
- ➤ accounting concepts and financial statements
- ➤ greatest challenges
- ➤ important techniques.

✓ What you need to know

Preparing financial statements for a sole trader requires an ability to construct:

- ➤ an income statement and
- ➤ a statement of financial position working.

Information required will usually be given in the form of:

- ➤ a trial balance
- ➤ a list of account balances

accompanied by notes detailing adjustments which need to be made to ensure compliance with accounting concepts.

4.1 Preparing an income statement

An income statement should be set out in a vertical format with each item correctly described and significant subtotals clearly labelled:

- ➤ Cost of sales
- ➤ Gross profit
- ➤ Profit (or loss) for the year.

The statement should be correctly titled, e.g. Income statement for the year ended 31 December 2016.

4.2 Preparing a statement of financial position

Statements of financial position should be set out to show subheadings and subtotals and totals as follows:

- ➤ First part of the statement:
  - ➤ Non-current assets
  - ➤ Current assets
  - ➤ Total assets.
- ➤ Second part of the statement:
  - ➤ Capital account
  - ➤ Non-current liabilities
  - ➤ Current liabilities
  - ➤ Total capital and liabilities.

Key term

Income statement: a document which sets out a detailed calculation of a business's profit or loss for a financial period (usually a year).

💡 Remember

Brackets () should be used for negative items appearing in any column which is otherwise made up of positive figures.

Key term

Statement of financial position: a document which sets out details of a business's assets, capital and liabilities at a particular date (usually the end of a financial year).

AS Level

4.3 Accounting concepts and financial statements

The following accounting concepts must be applied when preparing a sole trader's financial statement to ensure that the owner of the business and other users are given a true and fair view of the business's financial position.

➤ **Accruals (matching) concept:** requires costs and revenues to be matched for a financial period irrespective of the money received or paid. As a result it is necessary to make adjustments in financial statements for:

   ➤ Inventories: to ensure that only the cost of goods actually sold during a period are matched against the revenue for goods sold during that period.

   ➤ Depreciation: to ensure that the value of the benefit obtained from the use of non-current assets during a financial period is set against the revenue for that period.

   ➤ Expense adjustments: to ensure that the amount charged for expenses is based on the correct amount for the period irrespective of the amount paid, so that where amounts are due but unpaid, they are included in the charge to the income statement; and where amounts have been paid in advance for the next accounting period, they are excluded from the income statement.

   ➤ Income adjustments: where income is due for period but not yet received, it is included the income statement; where income has been received in advance for the next accounting period it is excluded from the income statement.

➤ **Realisation concept:** ensures that revenue is only accounted for in an income statement when there is certainty about the profit element, either because the customer has already paid for the goods or has promised to pay (i.e. an invoice for the goods has been issued).

➤ **Prudence concept:** requires losses to be recognised as soon as they are anticipated, but profits are ignored until they have been realised. This ensures that the owner of a business is not given an over-optimistic view of profits being made which could lead to poor decision making. Note that each of the following can also be seen to be a direct applicable to the accruals concept.

   ➤ Inventories: should be valued at cost or net realisable value whichever is lower. In other words inventories are valued on the basis of their original cost, or the estimated selling price less any the cost of any estimated expenses which will be incurred in order to put the goods into a saleable condition. The rule often has to be applied where goods have become damaged and cannot be sold at the normal selling price.

   ➤ Irrecoverable debts: as soon as it becomes apparent that the amount due from a credit customer will not be paid, the prudence concept requires the amount to be written off as an irrecoverable debt.

   ➤ Provision for doubtful debts: where there is uncertainty about the recovery of amounts due from trade receivables it is necessary to create a provision for doubtful debts to ensure the statement of financial position gives a true and fair view of the value of this current asset in accordance with the prudence concept.

- Other concepts also influence the way in which financial statements are prepared. The consistency principle, for example, is designed to ensure that accounting policies, such as depreciation methods, are applied in the same way each year to ensure results are comparable from one year to the next. The entity concept is applied by ensuring that the owner's private transactions do not affect the calculation of profit, so drawings, for example, is excluded from the income statement. The money measurement concept means that any important characteristics of a business which cannot be measured in monetary terms do not appear in the financial statements (e.g. good location, management expertise, etc.).

4.4 What are the greatest challenges?

- Assimilating a lot of detailed information required to present the financial statements. It is all too easy to overlook an item of information resulting in an incorrect figure or the totals of the statement of financial position failing to agree.

- Correctly treating each adjustment to be made when preparing the income statement, ensuring that the account is taken of all the details provided.

- Correctly recording the impact of adjustments when preparing the statement of financial position.

- Presenting the statements well under exam pressure.

4.5 Review some important techniques

Making adjustments to expense and income items: the table below summarises the effect of these adjustments in financial statements.

▼ **Table 4.1** Making adjustments in financial statements

Item	Adjustment required at year end	Effect on amount shown in trial balance	How recorded in statement of financial position
Expense	Accrual	Decreased	Current liability
Expense	Prepayment	Increased	Current asset
Income	Due	Increased	Current asset
Income	Received in advance	Decreased	Current liability

▼ **Example 4.1** Calculating the amount of an adjustment

Assume financial statements are being prepared for the year ended 31 December 2016.

A business's trial balance (debit column) included the item: Rent $1800. A note in the additional information states that rent $480 for the three months ending 28 February 2017 is due but unpaid.

Rent to be shown in the income statement needs to be increased by the amount due. However, in accordance with the accruals concept, only one month out of the three months to 28 February 2017 relates to the financial year under review. So the adjustment will be for one-third of $480 (i.e. $160) and the correct charge for rent for the year ended 31 December 2016 will be $1960.

> **Adjusting the provision for doubtful debts**
>
> Where a business has set up a provision for doubtful debts at some point in the past, it will be necessary to review the amount of the provision at each year ended to ensure it keeps in step with the total of trade receivables.

▼ **Example 4.2** Adjusting a provision for doubtful debts

A business's accounting year ends on 31 December. In 2014 a provision for doubtful debts was created at 5 per cent of trade receivables; this amounted to $460. At 31 December 2015 trade receivables totalled $9900; at 31 December 2016 trade receivables totalled $8800. The business's policy is to maintain the provision for doubtful debts at 5 per cent of trade receivables at the year end.

Action required:

▼ **Table 4.2**

Year ended 31 December	Provision for doubtful debts should be	Entry required income statement	Entry required in statement of financial position
2015	5% × 9900 = $495	Increase in provision $35 (i.e. profit decreased $35)	Trade receivables $9900 less provision for doubtful debts $495 = net $9405
2016	5% × $8800 = $440	Decrease in provision $55 (i.e. profit increased $55)	Trade receivables $8800 less provision for doubtful debts $440 = net $8360

Recording adjustments in the journal and ledger accounts

Here are some examples of how adjustments are recorded in the double entry system.

▼ **Example 4.3** Recording adjustments

A business's trial balance at 31 December 2016 (the accounting year-end) included the following details:

	Dr $	Cr $
General expenses	13 450	
Insurance	9 870	
Provision for doubtful debts		780
Rent receivable		6 540

Additional information at 31 December 2016:

➤ General expenses $390 was due but unpaid.

➤ An insurance premium of $720 for the six months ending 30 April 2017 had been paid in advance.

➤ A tenant owed rent of $440; another tenant had paid rent of $1080 in advance for the three months ending 31 January 2017.

➤ The provision for doubtful debts should be reduced by $110.

The journal and account entries to record the adjustments are as follows:
(Narratives in the journal entries and dates have been ignored.)

GENERAL JOURNAL

	$	$
Income statement	13 840	
General expenses		13 840
Income statement	9 390	
Insurance		9 390
Rent received	6 620	
Income statement		6 620
Provision for doubtful debts	110	
Income statement		110

Insurance account

Dr				Cr
Balance	9 870	Income stmnt	9 390
		Balance c/d	480
	9 870		9 870
Balance b/d	480		

Rent received account

Dr				Cr
Income stmnt	6 620	Balance	6 540
Balance c/d	360	Balance c/d	440
	6 980		6 980
Balance b/d	440	Balance b/d	360

General expenses account

Dr				Cr
Balance	13 450	Income stmnt	13 840
Balance c/d	390		
	13 840		13 840
		Balance b/d	390

Provision for doubtful debts account

Dr				Cr
Income stmnt	110	Balance	780
Balance c/d	670		
	780		780
		Balance b/d	670

Revision checklist

I can:

➤ prepare well-presented financial statements with correct headings, correct labelling of subtotals, correct use of brackets for negative items, correct layouts ☐

➤ make adjustments to expense and income items, depreciation charges and for provisions for doubtful debts, including making entries in the general journal and ledger accounts. ☐

Raise your grade

Seeta has extracted the following trial balance from her books of account at 31 December 2016:

	Dr $	Cr $		$	$
Capital		51 770	Provision for doubtful debts		380
Carriage inwards	580		Purchases	84 560	
Cash at bank	4 940		Rent received		6 200
Drawings	25 850		Returns inwards	1 290	
Insurance	3 960		Returns outwards		810
Inventory, 1 January 2016	11 420		Revenue		147 330
Irrecoverable debts written off	170		Trade payables		6 770
Non-current assets			Trade receivables	8 400	
cost	72 000		Wages and salaries	21 490	
provision for depreciation		21 400		234 660	234 660

Additional information at 31 December 2016:

- Inventory was valued at $13 550.

- Wages $290 were due but unpaid.

- An insurance premium of $640 had been paid to provide cover for the four months ending 31 March 2017.

- Rent receivable $510 was due but not received.

- The provision for doubtful debts was to be maintained at 5 per cent of trade receivables.

- Depreciation is provided at 20 per cent per annum using the reducing balance method.

Prepare an income statement for the year ended 31 December 2016 and a statement of financial position at that date.

Student answer

Seeta

Income statement for y/e ended 31 Dec 2016 ①

	$	$	$
Revenue		147 330	
less Returns inwards		1 290 ①	
			146 040
Opening inventory		11 420	
Purchases	84 560		
Add carriage inwards ①	580		
	85 140		
Less returns outwards ①	(810)		
		84 330	
		95 750	
Closing inventory		13 550 ①	
②			(82 200)
Gross profit			63 840
Add Rent received			6 710
			70 550
Less irrecoverable debts written off		170	
Insurance ③		3 320	
PDD ①		420	
Wages and salaries		21 780	
Depreciation of non-current assets ④		10 120	
			35 810 ①
②			34 740

SFP 31 Dec 2016 ①

	$	$	$
Non-current assets			
cost		72 000	
less prov for deprcn ①		31 520 ①	
			40 480
Current assets			
Inventory		13 550	
Trade receivables	8 400		
less PDD ①	420		
		7 980	
Other receivables		640	
Cash at bank		4 940	
			27 110
②			67 590
Capital			
Opening balance		51 770	
Profit for year		34 740	
		86 510	
Less drawings		25 850 ①	
			60 660
Current liabilities			
Trade payables		6 770	
Other payables ($290 + 510 ⑤)		800	
			7 570
②			68 230

How to improve this answer

① Poor presentation including: use of abbreviations in headings and for details in the financial statements, inconsistent use of brackets for negative items; carriage inwards incorrectly recorded before returns outwards in the income statement.

② Some key labels omitted: cost of sales, profit for the year, total assets, capital, total capital and liabilities.

③ Incorrect treatment of the adjustment to insurance.

④ Failure to provide workings for a more complex calculation (depreciation).

⑤ Incorrect treatment of an adjustment in the statement of financial position.

Model answer

Seeta

Income statement for the year ended 31 December 2016

	$	$	$
Revenue		147330	
less Returns inwards		(1290)	
			146040
Opening inventory		11420	
Purchases	84560		
less Returns outwards	(810)		
	83750		
Add Carriage inwards	580		
		84330	
		95750	
Closing inventory		(13550)	
Cost of sales			(82200)
Gross profit			63840
Add Rent received ($6200 + 510)			6710
			70550
Less irrecoverable debts written off		170	
Insurance $3960 – (3/4 x $640)		3480	
Increase in provision for doubtful debts			
(5% x $8400) – $380		40	
Wages and salaries ($21490 + 290)		21780	
Depreciation of non-current assets W1		10120	
			(35590)
Profit for the year			34960

W1: Depreciation charge

20% × (cost $72000 – provision $21400) i.e. 20% × $50600 = $10120

The correct answer is shown below:

Statement of financial position at 31 December 2016

	$	$	$
Non-current assets			
Cost		72 000	
Less provision for depreciation		(31 520)	
			40 480
Current assets			
Inventory		13 550	
Trade receivables	8 400		
Less provision for doubtful debts	(420)		
		7 980	
Other receivables ($510 + $480)		990	
Cash at bank		4 940	
			27 460
Total assets			67 940
Capital			
Opening balance		51 770	
Profit for year		34 960	
		86 730	
Less drawings		(25 850)	
			60 880
Current liabilities			
Trade payables		6 770	
Other payables		290	
			7 060
Total capital and liabilities			67 940

★ **Exam tips**

Key features of the model answer

➤ Careful attention to matters of details: avoidance of the use of abbreviations, brackets used to indicate negative amounts within a column, correct order of returns outwards before carriage inwards.

➤ All subtotals labelled in the income statement and statement of financial position.

➤ Insurance adjustment takes account of all the information given and some workings are shown.

➤ Workings for the depreciation calculation are included.

➤ The adjustments are correctly recorded in the statement of financial position (i.e. the recent received due is now shown as a current asset).

❓ Exam-style questions

1 Describe the entries to be made when decreasing a provision for doubtful debts.

2 Explain how the accruals concept is applied when preparing an income statement.

3 The following trial balance was extracted from the books of the business owned by Kisha:

	Dr $	Cr $
Bank overdraft		3120
Bank loan (repayable 2017)		6400
Bank loan interest	430	
Capital		145300
Carriage inwards	510	
Carriage outwards	220	
Discounts received		420
Drawings	29400	
Furniture and fittings		
Cost	24300	
Provision for depreciation		7700
General expenses	4730	
Irrecoverable debts written off	90	
Inventory, 1 January 2016	11950	
Premises		
Cost	190000	
Provision for depreciation		15200
Provision for doubtful debts		160
Purchases	181300	
Rent received		7300
Revenue		292860
Returns inwards	3370	
Returns outwards		1920
Staff wages	36400	
Trade payables		12360
Trade receivables	9200	
Water charges	840	
	492740	492740

Additional information at 31 December 2016:

- Inventory was valued at cost $8470. This included some items which had cost $760 but which now have a net realisable value of $410.

- General expenses $310 was prepaid.

- Loan interest at 5 per cent per annum was due for the last three months of 2016 on the bank loan.

- Rent received of $520 was received in advance.

- Depreciation should be charged at 20 per cent per annum on furniture and fittings using the reducing balance method, and at 2 per cent per annum on premises using the straight-line method.

- The provision for doubtful debts should be maintained at 2.5 per cent of trade receivables.

Prepare the business's income statement of the year ended 31 December 2016 and a statement of financial position at that date.

Key topics

➤ calculating profits/losses where there is minimal information

➤ assessing profits/losses where it is possible to construct an income statement

➤ adjustments for expense and income items

➤ using ratios to find missing figures

➤ accounting concepts and incomplete records

➤ greatest challenges

➤ important techniques.

✓ What you need to know

The owners of many smaller businesses do not keep full accounting systems often through lack of time and lack of expertise. However, it is important that they can assess profit or losses for each financial period in order to meet the requirements of tax authorities and also to ensure they have adequate information about their businesses in order to support management and decision making. There are two different techniques for assessing profits or losses for a period: using statements of affairs or preparing income statements.

5.1 Calculating profits/losses where there is minimal information

This technique requires the drawing together of information about the capital of a business in order to find the one missing element: the profit or loss for the period. For this, it is necessary to know:

➤ The opening capital for the period: which can be found by using the basic accounting formula (assets = liabilities + capital) and preparing a statement of affairs. This resembles a statement of financial position, but differs from the latter because it cannot be constructed from balances in ledger accounts.

➤ The closing capital for the period which can be found in the same way.

➤ The drawings and any additions to capital.

Key term

Statement of affairs: a list of the assets, liabilities and capital of a business which does not maintain full accounting records.

5.2 Assessing profits/losses where it is possible to construct an income statement

If, in addition to the minimal information describes above, the owner(s) of the business has maintained a cash book, it will be possible using a variety of techniques to construct an income statement. This arrangement is sometimes referred to as a single entry accounting system. The techniques required to prepare the income statement are as follows:

Credit sales

Credit sales are calculated by reconstructing a total trade receivables account (in effect a sales ledger control account). In the most straightforward situations this means including:

➤ total opening and total closing balances

➤ total receipts from credit customers.

However, all transactions affecting trade receivables must be included so the total account may also need to include:

➤ total returns inwards

➤ total discounts allowed

➤ total irrecoverable debts written off.

Credit purchases

A similar process is used to find total credit purchases and a total trade payable account is prepared. The total account will include:

➤ total opening and total closing balances

➤ total payments to credit supplies.

And possibly:

➤ total returns outwards

➤ total discounts received.

5.3 Adjustments for expense and income items

It may be necessary to make the usual adjustments to income items (for amounts due or amounts received in advance) or to expense items (for amounts due or prepaid) at the year end. In addition there could be information about adjustments at the *beginning* of the period under review. Here is a summary of how adjustments should be treated:

Item	Adjustment	Treatment	Reason
Expense	Opening balance due	**Exclude**	The amount due relates to the previous year.
	Opening balance prepaid	**Include**	The amount prepaid relates to the current year.
	Closing balance due	**Include**	The amount due relates to the current year.
	Closing balance prepaid	**Exclude**	The amount prepaid relates to the next year.
Income	Opening balance received in advance	**Include**	The amount received in advance is for the current year.
	Opening balance due	**Exclude**	The amount due relates to the previous year.
	Closing balance received in advance	**Exclude**	The amount received in advance relates to the next year.
	Closing balance due	**Include**	The amount due relates to the current year.

> ★ **Exam tip**
>
> The pattern which emerges for making adjustments can be summarised as:
>
> ➤ amount relates to the current year: **Include**
>
> ➤ amount relates to the previous year or next year: **Exclude**.

Depreciation charges

Depreciation charges may have to be calculated by comparing the value of a non-current asset at the end of the financial period with its value at the beginning of the financial period. It may, of course, be necessary to take account of the value of any additions made during the year and any disposals.

5.4 Using ratios to find missing figures

Where a business has a known mark-up policy (or 'gross profit margin') or a known rate of inventory turnover, the ratios can be used to find missing information in the trading section of the income statement. Quite often, the missing figure is that of inventory lost or stolen during the year, but ratios could be used to find revenue or purchases. The technique is to assemble the trading section of the income statement with all the known information and then to use the ratio given to find the one missing item. It is useful to remember how to convert a mark-up policy to the ratio for gross profit margin (or vice versa) to facilitate calculations in some cases. Here are two examples using a very simple approach:

Converting mark-up to gross profit margin		Converting gross profit margin to mark up	
Example 1	**Example 2**	**Example 3**	**Example 4**
Mark up is 25% (i.e. 1/4)	Mark up is 60% (i.e. 3/5)	Gross profit margin is 331/3% (i.e. 1/3)	Gross profit margin is 62.5% (i.e. 5/8)
Assemble a 'skeleton' trading section to show the ratio details in the simplest form			
Revenue — / Cost of sales 4 / Gross profit 1	Revenue — / Cost of sales 5 / Gross profit 3	Revenue 3 / Cost of sales — / Gross profit 1	Revenue 8 / Cost of sales — / Gross profit 5
Complete the missing element in the skeleton trading section (shown here in bold)			
Revenue **⑤** / Cost of sales 4 / Gross profit 1	Revenue **⑧** / Cost of sales 5 / Gross profit 3	Revenue 3 / Cost of sales **②** / Gross profit 1	Revenue 8 / Cost of sales **③** / Gross profit 5
The gross profit margin is 1/5, i.e. 20%	The gross profit margin is 3/8, i.e. 37.5%	The mark up is ½, i.e. 50%	The mark up is 5/3 i.e. 166.67%

Missing cash

Where cash is missing or stolen (or perhaps used for unrecorded drawings) it will be necessary to prepare a cash account to include all the known cash transactions. The 'balancing' item will be the missing cash.

5.5 Accounting concepts and incomplete records

As the main consideration is preparing financial statements the focus is on applying the accruals, prudence and consistency concepts (see Unit 4 for more details).

5.6 What are the greatest challenges?

➤ Assimilating a considerable amount of detail and appreciating what figures need to be calculated.

➤ Providing detailed workings for missing figures where the calculations are complex.

➤ Using ratios to find missing figures.

Note: preparing a totally accurate set of financial statements from incomplete records is one of the most skilful activities in financial accounting.

5.7 Review some important techniques

Calculating profit/loss from minimal information

▼ **Example 5.1**

The owner of a business provides the following information about the year ended 31 October 2016:

	$
Capital, 1 November 2015	47 900
Drawings during the year ended 30 October 2016	14 850
Additional capital introduced May 2016	8 500
Capital, 31 October 2016	38 650

The answer is:

	$
Decrease in capital during the year	(9 250)
Add back drawings	14 850
Deduct capital introduced	(8 500)
So, loss for the year is	(2 900)

> **✗ Common error**
>
> To deduct drawings (rather than add them back) or to add extra capital introduced (rather than deduct).

Calculating revenue

▼ **Example 5.2**

The owner of a business wishes to know the total value of sales made during the year ended 30 September 2016. The following information is available.

	$
Amounts due to credit customers at	
1 October 2015	7 490
30 September 2016	6 840
Receipts from credit customers	92 310
Discounts allowed to credit customers	880
Returns inwards	1 480

In addition, cash taking banked totalled $112 300 and cash takings were used to pay staff wages $14 730 and for drawings $4820.

Total revenue for the year ended 30 September 2016 is calculated as follows:

Dr		Sales ledger control account		Cr
	$			$
Opening balance	7 490	Receipts		92 310
Credit sales	94 020	Discounts allowed		880
		Returns inwards		1 480
		Closing balance c/d		6 840
	101 510			101 510
Balance b/d	6 840			

	$	$
Credit sales		94 020
Cash sales		
Takings banked	112 300	
Takings not banked		
Staff wages	14 730	
Drawings	4 820	
		131 850
Total revenue		225 870

Calculating depreciation

▼ **Example 5.3** The following details are available about a business's motor vehicles

	$
Net book value at	
1 January 2016	43 600
31 December 2016	44 100
Additions purchased during 2016	14 400

During the year a motor vehicle which had cost $17 000 and on which depreciation of $12 100 had been charged was disposed of for $3500.

The depreciation charge for 2016 is:

	$	$
Net book value at 1 January 2016	43 600	
Additions	14 400	
	58 000	
Less net book value of disposal		
(17 000 – 12 100)	4 900	
		53 100
Less net book value at 31 December 2016		44 100
Depreciation		9 000

Example 5.4 Calculating value of missing inventory

Fire occurred in a business's storerooms and inventory was destroyed. The owner of the business is able to supply the following details for the year ended 31 December revenue $240,000, purchases $189,600, inventory 1 January 2016 $17,500, inventory 31 Deember2016 $14,500.

	$
Revenue	240 000
Purchases	189 600
Inventory at	
1 January 2016	17 500
31 December 2016	14 500

The business's policy is to achieve a mark-up of 25 per cent on all goods sold.

The trading section of the income statement is shown below with all information shown except the value of the missing inventory.

	$	$	$
Revenue			240 000
Opening inventory		17 500	
Purchases		189 600	
		207 100	
Closing inventory			
Actual	14 500		
Destroyed in fire	?? ???		
		15 100	
Cost of sales			192 000
Gross profit (25% of cost of sales = 20% of revenue)			48 000

So, the inventory destroyed in the fire is the missing figure of $600.

> **Remember**
>
> If the loss of inventory is not covered by insurance, it must be entered in the list of business's expenses. If, however, it is covered by insurance, the amount owed by the insurance company will appear as a current asset in the statement of financial position.

▼ **Example 5.5** Calculate value of missing cash

The owner of a business believes a dishonest employee may have stolen cash. The following information about cash transactions is available:

	$
Balance of cash in hand 1 January 2016	495
Cash takings	34 724
Staff wages	11 348
Drawings	7 472
Cash takings banked	15 448
Balance of cash in hand 31 December 2016	337

The cash account based on this information is shown below. The missing amount is the cash apparently stolen by an employee.

Dr	Cash account		Cr
	$		$
Opening balance	495	Staff wages	11 348
Cash takings	34 724	Drawings	7 472
		Cash takings banked	15 448
		Missing cash	**614**
		Balance c/d	337
	35 219		35 219
Balance b/d	337		

I can:

➤ calculate a business's profit or loss from minimal information ☐

➤ calculate credit sales and credit purchases using control accounts ☐

➤ calculate expense and income items taking account of opening and closing adjustments ☐

➤ calculate depreciation charges comparing opening and closing net book values of non-current assets ☐

➤ apply ratios to the calculation of missing information from the trading section of an income statement ☐

➤ calculate missing cash using a cash account ☐

➤ prepare detailed workings to support information shown in the final answer. ☐

 Raise your grade

Sam is a sole trader. He does not maintain full accounting records. He has provided the following information for the year ended 31 December 2016.

	1 January 2016	31 December 2016
	$	$
Administration expenses prepaid	370	
Administration expenses due		410
Equipment at valuation	14930	16850
Interest receivable due		180
Inventory	13360	?
Trade payables	11920	13770
Trade receivables	5660	5030
Investment: loan to J Patel	8000	8000

Summary of Sam's bank account for the year:

Receipts	$	Payments	$
Opening balance	2990	Administration expenses	6840
Receipts from credit customers	62340	New equipment	8200
Cash sales	48490	Payments to trade suppliers	81270
Interest received	300	Drawings	18810
Proceeds of sale of equipment	1720	Closing balance c/d	720
	115840		115840
Balance b/d	720		

Additional information:

• All cash sales had been banked with the exception of payments made for staff wages $15390.

• Credit customers were allowed cash discounts of $410.

• Discounts of $780 were received from trade suppliers.

• Sam took goods for his own use during the year valued at $350.

- Equipment sold during the year had a net book value of $1500.
- All goods are sold with a mark-up of 50 per cent.

Prepare an income statement for the year ended 31 December 2016.

Student answer

W1   Revenue

Sales ledger control account

Opening balance	5660	Receipts ❶	62340
Credit sales	61710	Closing balance c/d	5030
	67370		67370

	$
Cash sales banked	48490
Cash sales not banked	15390
Credit sales	61710
	125590

W2   Purchases

Purchases ledger control account

Payments	81270	Opening balance	11920
Discounts received	780	Credit purchases ❷	83900
Closing balance c/d	13770		
	95820		95820
		Balance b/d	13770

W3   Administration expenses

	$
Payment	6840
Less opening prepaid ❸	370
Add closing due	410
	6880

W4   Equipment depreciation

	$
Opening net book value	14930
Additions	8200
❹	23130
Closing net book value	(16850)
Depreciation	6280

Sam

Income statement for the year ended 31 December 2016

	$	$
Revenue (W1) ❺		125590
Opening inventory	13360	
Purchases (W2) ❻	83900	
	97260	

Closing inventory ❼	(34 465)	
Cost of sales		(62 795)
Gross profit (1/2 revenue) ❼		62 795
Interest received ($300 + 180) ❽		480
		63 275
Discounts allowed	410	
Administration expenses (W3)	6 880	
Depreciation of equipment (W4)	6 280	
❾		(13 570)
Profit for year		49 705

How to improve this answer

① The candidate has forgotten to include the discounts allowed in the calculation of credit sales.

② The calculation of credit purchases does not take account of goods for own use.

③ The adjustment for the opening prepayment of administration expenses should have been added.

④ The calculation of depreciation is incorrect because the disposal of the non-current asset has been ignored.

⑤ The calculation of revenue was incorrect.

⑥ The calculation of purchases was incorrect.

⑦ The closing inventory has been miscalculated because the treatment of the mark-up ratio is incorrect.

⑧ The list of income items is incomplete: discounts received and the profit on disposal have been overlooked.

⑨ The list of expenses items is incomplete: wages have been overlooked.

Model answer

W1 Revenue

Sales ledger control account

Opening balance	5 660	Receipts	62 340
Credit sales	62 120	Discounts allowed	410
		Closing balance c/d	5 030
	67 780		67 780
Balance b/d	5 030		

Cash sales banked	48 490
Cash sales not banked	15 390
Credit sales	62 120
	126 000

W2 Purchases

Purchases ledger control account

Payments	81 270	Opening balance	11 920
Discounts received	780	Credit purchases	83 900
Closing balance c/d	13 770		
	95 820		95 820
		Balance b/d	13 370

Net purchases: $83 900 less goods own use $350 = $83 550

W3   Administration expenses

	$
Payment	6840
Add opening prepaid	370
Add closing due	410
	7620

W4   Equipment depreciation

	$
Opening net book value	14930
Additions	8200
	23130
Disposal net book value	(1500)
	21630
Closing net book value	(16850)
Depreciation	4780

W5   Profit/loss on disposal

	$
Proceeds	1720
less net book value	1500
Profit	220

Sam

Income statement for the year ended 31 December 2016

	$	$
Revenue (**W1**)		126000
Opening inventory	13360	
Purchases (net) (**W2**)	83550	
	96910	
Closing inventory	(12910)	
Cost of sales		(84000)
Gross profit (1/2 cost of sales, i.e. 1/3 revenue)		42000
Discount received	780	
Profit on disposal of equipment (**W5**)	220	
Interest received ($300 + 180)	480	
		1480
		43480
Discounts allowed	410	
Administration expenses (**W3**)	7620	
Depreciation of equipment (**W4**)	4780	
Wages	15390	
		(28200)
Profit for year		15280

★ Exam tips

Key features of the model answer

➤ Full workings have been included (the original answer omitted the calculation of the profit on disposal).

➤ The workings take account of all the information given (i.e. discounts allowed, discounts received, goods for own use, asset disposal net book value, etc.).

➤ The mark-up ratio has been applied correctly to the calculation of gross profit and closing inventory.

➤ The income and expense sections include all relevant information.

1 Define the term 'statement of affairs'.

2 Explain what is meant by 'single entry accounting'.

3 A sole trader recently opened a business. The owner does not plan to keep full accounting records. Discuss the owner's decision not to keep full accounting records.

4 Anais is a sole trader. She does not maintain full accounting records. She has provided the following information.

Statement of financial position at
31 December 2016

	$	$
Non-current assets at net book value		84500
Current assets		
Inventory	17490	
Trade receivables	13410	
Other receivables (rent receivable)	490	
Cash at bank	1100	
Cash in hand	730	
		33220
Total assets		117720
Capital		105230
Current liabilities		
Trade payables	11730	
Other payables (operating expenses)	760	
		12490
Capital and liabilities		117720

Additional information for the year ended
31 December 2016:

	$
Cash takings	46380
Cheques received from credit customers	179450
Discounts received	1180
Operating expenses paid by cheque	77510
Payments to credit suppliers by cheque	120840
Proceeds from disposal of non-current assets (banked)	3800
Rent received (banked)	11220
Returns inwards	540

Cash takings of $21000 were banked. Cash $2700 was used to purchase goods for resale. Cash was also taken by the owner of the business for personal use, but no record was kept of the total amount taken.

The non-current asset disposed of during 2016 had a net book value of $4500. There were no purchases of non-current assets during 2016.

On 31 December 2016 the business's assets and liabilities were:

	$
Non-current assets	71200
Inventory	14160
Trade receivables	11100
Other receivables (operating expenses prepaid)	520
Cash at bank	18220
Cash in hand	1480
Trade payables	13770
Other payables (rent received in advance)	830

Prepare an income statement for the year ended 31 December 2016 and a statement of financial position at 31 December 2016.

Partnership accounts

AS 1.4.3A, 1.4.3B SB pages 140–173

Key topics

➤ partnership accounts

➤ changes in partnership agreements

➤ accounting concepts

➤ greatest challenges

➤ important techniques.

✓ What you need to know

Partnership accounts involve: the preparation of capital and current accounts of partners; appropriation accounts showing the division of profits and losses including where agreements have changed during a financial period; goodwill adjustments; the retirement of a partner; the admission of a partner; the dissolution of a partnership.

6.1 Partnership accounts

Partnerships are often formed by a group of individuals to raise a larger amount of capital than any one individual could provide. Individuals might also bring specialist skills and expertise to a business; and the management of a business could be less onerous with the opportunity to share responsibilities. However, decision making can be slower as the agreement of all the partners will be required to make major changes. There is also the chance that disagreements and disputes between partners could seriously affect a business's performance. Of course, all the profit of the business must now be shared between the partners. Just like a sole trader, each partner has unlimited liability for the debts of the business.

Accounting records

Usually partners maintain separate accounts for:

➤ capital contributions (fixed capital accounts)

➤ shares of profits/losses (current accounts)

➤ drawings (the totals of which are transferred to the current accounts at the year-end).

Occasionally, partners may agree to use a combination of the capital, current and drawings accounts, referred to as fluctuating capital accounts.

Profit/loss sharing

Partners normally make a formal agreement about how profits and losses should be shared and about other matters concerning the finances and management of the business (for example, limits on drawings, amounts of fixed capitals, individual responsibilities, etc.). In written form an agreement is called a 'Deed of Partnership'.

> **Key term**
>
> **Current accounts**: accounts for recording the changes in a partner's contribution arising from drawings and shares of profits or losses.

Profit and loss sharing can take a variety of forms, for example an agreement to:

- share profits and losses equally or in some other ratio
- share profits and losses after providing for:
  - interest on drawings (which penalises partners according to the amount of their drawings during a financial period)
  - interest on capitals (which rewards partners according to the size of their fixed capitals)
  - salaries (which rewards an individual partner(s) for specific contributions to the running of the business).

Details are set out in an appropriation account.

> **Remember**
>
> Interest on drawings should be added to the profit for the year within the appropriation account, unlike other agreed ways of sharing profits (such as interest on capital and salaries) which are deducted.

Loans from partners

Where a partner makes a loan to the business any agreed interest is charged as an expense to the income statement.

Partnerships with no agreement

Where partners have not made any formal agreement about sharing profits and losses, and where they are in dispute about how profits and losses should be shared, the Partnership Act 1890 provides that profits and losses should be shared equally and no other appropriations of profit should be permitted (such as interest on capital, etc.). Where a partner has made a loan to the business, interest at 5 per cent per annum should be charged to the income statement.

6.2 Changes in partnership agreements

Partners are free to make changes to their agreement at any time during a financial period. Where this occurs it is necessary to divide the profit or loss for the year proportionately between the times in which each agreement was in operation. Care is required to ensure that those aspects of each agreement which are for a time period are correctly apportioned. For example, if 9 months into a financial year, partners decided to introduce interest on capital at 10 per cent per annum on their fixed capitals, it will be necessary to calculate interest on capital for just the last three months of the year in which the new agreement is introduced.

Changes in ownership

A change in ownership may arise because:

- a partner leaves the partnership – possibly due to retirement
- a new partner joins the partnership – possibly to replace a retiring partner, or to provide additional skills, expertise and/or capital.

When a change occurs it is important for the partners to ensure the balances of their capitals accounts properly reflects the current value of the partnership. This usually necessitates a:

- revaluation of tangible assets
- valuation being placed on the intangible asset, goodwill.

Where the revaluation of tangible assets results in a surplus, the capital accounts of the partners are credited; where the result is a deficit, the capital accounts of the partners are debited. Surpluses/deficits are shared in accordance with the partners' profit/loss sharing agreement.

In the case of goodwill, it is usual not to maintain a goodwill account in the books of the partnership. The effect of any valuation of goodwill is to benefit the original partners who have been responsible for the success of the business and who have created the goodwill, so these partners see a net increase in their capital balances.

Financing the amount due to a retiring partner

Is one of the problems facing the partners when a retirement takes place. The amount due to the retiring partner can be financed in any one of a number of ways agreed by all the partners:

➤ leaving the balance owing to the retiring partner on loan (or part of the amount owing)

➤ remaining partners introducing extra capital in order to pay off amount due

➤ admitting a new partner(s) to replace the retiring partner

➤ borrowing the funds required from an outside source, e.g. bank loan.

Dissolution of partnership

A partnership may cease to exist where the business is making losses, where a partner retires or a partner dies, or where the partners cannot resolve a dispute. In a dissolution, assets are sold or taken over by a partner at an agree valuation, and liabilities are discharged. Details are recorded in a realisation account which is designed to show a profit or loss on the dissolution which is then shared between the partners in accordance with their profit/loss agreement. The last stage in the dissolution process is to settle the final balances on capital accounts by transferring funds from (or to) the bank account.

6.3 Accounting concepts

Of special interest in partnership accounts is the application of the concept of historical cost in relation to goodwill. When partners decide to place a value on goodwill they are in effect giving their valuation of an internally-generated asset rather than a purchased asset. The historical cost concept requires only transactions with an actual cost of purchas to be recorded to ensure objectivity. There can be no objective valuation of the asset which is why it is usual practice not to record goodwill in an account, but merely to make an adjustment for its value.

> ⭐ **Exam tip**
>
> When recording the retirement of a partner do not included goodwill in the revaluation account. Instead make a separate record of adjustments for this item.

> **Key term**
>
> **Realisation account:** an account for recording the gains and losses arising from the disposal of assets and settlement of liabilities upon the dissolution of a partnership.

> 💡 **Remember**
>
> When closing the bank account, remember to settle the accounts of each partner based on the final balance of their accounts. It is a common mistake to try to divide the remaining bank balance in the partners' profit sharing ratio.

6.4 What are the greatest challenges?

➤ Implementing the exact terms of a partnership agreement.

➤ Preparing accurate records of the appropriation of profits/losses in the financial statement and current accounts, particularly where an agreement has changed during a financial year.

➤ Correctly recording the retirement of a partner including the revaluation of assets and an adjustment for goodwill.

➤ Correctly recording the admission of a new partner including an adjustment for goodwill.

➤ Recording the dissolution of a partnership, including the calculation of the realisation profit or loss, and the final settlement of the amounts due to (or by) each partner.

6.5 Review some important techniques

Preparing current accounts

The format for partners' current accounts:

 Exam tip

If a question asks for current accounts, do not present the details in a vertical statement and risk losing many marks.

Current accounts

Dr	Partner	Partner		Partner	Partner	Cr
	X	Y		X	Y	
	$	$		$	$	
Opening balance	xx	xx	Opening balance	xx	xx	
Drawings	xx	xx	Interest on capitals	xx	xx	
Interest on drawings	xx	xx	Salary	xx	xx	
Share of losses	xx	xx	Share of profits	xx	xx	
Balances c/d	xx	xx	Balances c/d	xx	xx	
	xx	xx		xx	xx	
Balances b/d	xx	xx	Balances b/d	xx	xx	

It is usually considered good practice to set out current accounts in columnar form as this is a more efficient use of time. A partner may have a:

➤ credit balance (indicating that the partner's profits shares have exceeded drawings) so the business is in debt to the partner

➤ a debit balance (indicating that drawings have exceeded profit shares) so the partner is in debt to the business.

Preparing an appropriation account where there is a change

in the agreement

▼ **Example 6.1** Eve and Frank have been in partnership for a number of years

Their fixed capitals are Eve $140 000, Frank $80 000. They have been sharing profits and losses equally after allowing interest on capital at 10 per cent per annum. Their financial year ends on 31 December. On 1 October 2016, the partners decided to change their profit/loss sharing agreement. The terms of the new agreement are:

➤ interest on capitals to be allowed at 10 per cent per annum

➤ Frank to be allowed a salary of $16 000 per annum

➤ profits and losses to be shared in the ratio Eve:Frank, 3:2.

Profit for the year ended 31 December 2016 was $120 000 and this was assumed to have accrued evenly throughout the year.

The appropriation account for the year ended 31 December 2016 is as follows:

		1 January–30 September		1 October–31 December	
		$	$	$	$
Profit for the year			90 000		30 000
Interest on capital	Eve	10 500		3 500	
	Frank	6 000		2 000	
			(16 500)		(5 500)
			73 500		24 500
Salary	Frank				(4 000)
					20 500
Profit shares	Eve	36 750		12 300	
	Frank	36 750		8 200	
			(73 500)		(20 500)

X Common error

To overlook the need to take account of time periods where an agreement changes. For example, Frank's partnership salary for 2016 applies to just the last three months of the year.

Goodwill

➤ Goodwill is an intangible asset (i.e. an asset with no physical presence) and is defined as the additional value of an established business, above the value of its tangible assets. In a partnership goodwill may arise because the business is successful as demonstrated by a number of factors such as:

  ➤ its ability to earn profits in the future

  ➤ sound management

  ➤ good labour and customer

  ➤ favourable location.

➤ Goodwill is difficult to value except where a business is sold for a value above its net book value as shown by a statement of financial position. Where there are changes in ownership in a partnership, the partners can adopt any one or a variety of methods to value the goodwill, for example: taking an average of the profits for the last few years.

➤ The value of goodwill is generally ignored, but when there is a change in ownership it is important that some adjustment is made for its value:

  ➤ **Retirement:** when a partner retires, the partners normally agree a value for goodwill and an adjustment is made in the partners' capital accounts to ensure that the retiring partner is duly rewarded.

  ➤ **Admission of a new partner:** all the partners agree a value for goodwill and an adjustment is made whereby the original partners benefit at the expense of the new partner.

▼ **Example 6.2** Goodwill adjustment on the retirement of a partner

Rachel, Steve and Tom have been in partnership for many years sharing profits and losses equally. Tom is due to retire from the partnership. Rachel and Steve will continue in business together and in future share profits and losses in the ratio 3:2 respectively. The partners have agreed that an adjustment should be made for their valuation of goodwill at $75 000.

The adjusting entries for goodwill in the partners' are:

Capital accounts

Dr					Cr		
	Rachel $	Steve $	Tom $		Rachel $	Steve $	Tom $
Goodwill adjustment	45 000	30 000		Goodwill adjustment	25 000	25 000	25 000

The net effect of the adjustment is that the retiring partner, Tom, benefits by $25 000, and the remaining partners see a net reduction in their capitals (Rachel $20 000, Steve $5 000). Of course, Rachel and Steve will have their own opportunity to benefit from the business's goodwill at some future date.

▼ **Example 6.3** Goodwill adjustment on the admission of a partner

Helen and Ian have been in partnership for a number of years with fixed capitals of Helen $160 000 and Ian $80 000; they have been sharing profits and losses in the ratio 7:3 respectively. They plan to expand their business and so they have invited Jane to join the partnership. In future the partners will share profits and losses in the ratio Helen:Ian:Jane, 3:1:1 respectively. Jane will provide additional capital of $150 000 consisting of a cheque for $120 000 and a vehicle worth $30 000. Helen, Ian and Jane have agreed that the goodwill of the partnership at the date of Jane's admission should be valued at $70 000 and that no goodwill account should be maintained in the partnership's books of account.

> **X Common error**
>
> When making a goodwill adjustment, to confuse who should benefit and who should not.

The capital accounts of the partners recording these arrangements are:

Capital accounts

Dr					Cr		
	Helen $	Ian $	Jane $		Helen $	Ian $	Jane $
Goodwill adjustment	42 000	14 000	14 000	Balances	160 000	80 000	
Balances c/d	167 000	87 000	136 000	Goodwill adjustment	49 000	21 000	
				Bank			120 000
				Vehicle			30 000
	209 000	101 000	150 000		209 000	101 000	150 000
				Balances b/d	167 000	87 000	136 000

In effect the original partners, Helen and Ian, benefit from the goodwill adjustment by $7000 each; the new partner, Jane, is penalised by $14 000.

Dissolution of partnership

The format for the realisation account is:

Realisation account

Dr			Cr
	$		$
Book value of each asset to be realised (i.e. sold or disposed of – normally all assets except the bank account)	xx xx xx xx xx xx	Amounts received as assets are disposed of:	
		Bank (for cash received)	xx
		Capital account (where a partner takes over an asset at an agreed value)	xx
Amounts paid to discharge liabilities	xx	Liabilities due to be discharged	xx
Other costs arising during dissolution	xx		
Balance (where there is a realisation **profit**) shared among the partners	xx xx xx	Balance (where there is a realisation **loss**) shared among the partners	xx xx xx
	xxx		xxx

The format for a partner's capital account in a dissolution is:

Capital account

Dr			Cr
	$		$
Current account (debit balance)	xx	Opening balance	xx
Assets taken over	xx	Current account (credit balance)	xx
Loss on realisation	xx	Profit on realisation	xx
Bank (where amount due to partner)	xx	Bank (where partner must pay off amount due to partnership)	xx
	xxx		xxx

I can:

➤ prepare partners' capital and current accounts recording relevant details correctly ☐

➤ explain the meaning of balances on partners' currents accounts ☐

➤ prepare an appropriation account correctly including the situation where an agreement has been changed during the year ☐

➤ prepare a revaluation account when a partner retires ☐

➤ explain what is meant by goodwill and how it is treated in the accounts of a partnership ☐

➤ make adjustments for goodwill on the retirement of a partner or upon the admission of a new partner ☐

➤ prepare a realisation account to show the profit or loss arising when a partnership is dissolved ☐

➤ make entries to close the accounts of a partnership when it is dissolved. ☐

↑ Raise your grade

Darim and Ellie have been in partnership sharing profits and losses in ratio 3:2. After a number of years of loss-making, the partners have decided to dissolve their partnership. Their statement of financial position on 31 October 2016 was as follows:

	$	$
Non-current assets		
Motor vehicles	49400	
Furniture and equipment	11940	
		61340
Current assets		
Inventory	9850	
Trade receivables	14200	
Bank	3320	
		27370
Total assets		88710
Capital accounts		
Darim	46000	
Ellie	32000	
		78000
Current accounts		
Darim	(3180)	
Ellie	2450	
		(730)
Current liabilities		
Trade payables		11440
Total capital and liabilities		88710

Additional information:

- The motor vehicles were sold for $32 000.

- Darim agreed to take over the equipment and furniture at an agreed valuation of $6800.

- Inventory was sold for $7100.

- A credit customer who owed $2200 could not be traced and the account was written off as irrecoverable. Remaining credit customers settled their accounts less cash discounts of 5 per cent.

- The accounts of trade payables were settled less discounts received of $460.

- The costs of dissolution were $1370.

Prepare a realisation account, the partners' capital accounts and the bank account recording the dissolution of the partnership.

Student answer

Realisation account

Dr				Cr
	$			$
Sundry assets		Bank		
Motor vehicles	49 400	Motor vehicles		32 000
Furniture and equipment	11 940	Inventory		7 100
Inventory	9 850	Trade receivables ❸		11 400
Trade receivables ❸	14 200	Capital, Darim ❷		
Bank ❶	3 320	Furniture and equipment		11 940
Irrecoverable debt	2 200	Trade payables ❹		11 440
Discounts allowed ❸	600	Discounts received ❹		460
Bank		Capitals (realisation loss)		
Trade payables ❹	10 980	Darim ❺		14 760
Dissolution costs	1 370	Ellie ❺		14 760
	103 860			103 860

Bank account

Dr			Cr
	$		$
Realisation ❻		Trade payables	10 980
Motor vehicles	32 000	Dissolution costs	1 370
Inventory	7 100	Capital accounts	
Trade receivables	11 400	Darim ❼	19 075
		Ellie ❼	19 075
	50 500		50 500

Capital accounts

Dr						Cr
		Darim	Ellie		Darim	Ellie
⑧		$	$		$	$
Realisation				Balances	46 000	32 000
Loss ⑤		14 760	14 760	⑧		
Bank ⑦		19 075	19 075		?	?
		33 835	33 835		46 000	32 000

How to improve this answer

① The bank account has been included in the assets to be realised.

② The value shown for the transfer of the motor vehicle to Darin should be the agreed transfer value not the net book value.

③ There is confusion over how to record the realisation of trade receivables.

④ There is confusion over how to record the settlement of trade payables.

⑤ The loss on dissolution has been divided among the partners in the wrong ratio.

⑥ The opening balance has been omitted from the bank account.

⑦ The amount to be paid to the partners has incorrectly been split in the profit sharing ratio instead of the remaining balance on the capital accounts.

⑧ The current account balances have been overlooked when compiling the capital accounts and the asset taken over by Darim has also been omitted from his capital account.

Model answer

Realisation account

Dr			Cr
	$		$
Sundry assets		Bank	
Motor vehicles	49 400	Motor vehicles	32 000
Furniture and equipment	11 940	Inventory	7 100
Inventory	9 850	Trade receivables	11 400
Trade receivables	14 200	Capital, Darim	
Bank		Furniture and equipment	6 800
Trade payables	10 980	Trade payables	11 440
Dissolution costs	1 370	Capital accounts (loss on	
		realisation)	
		Darim	17 400
		Ellie	11 600
	97 740		97 740

Bank account

Dr				Cr
	$			$
Opening balance	3 320	Realisation		
Realisation		Trade payables		10 980
Motor vehicles	32 000	Dissolution costs		1 370
Inventory	7 100	Capital accounts		
Trade receivables	11 400	Darim		18 620
		Ellie		22 850
	53 820			53 820

Capital accounts

Dr				Cr	
	Darim	Ellie		Darim	Ellie
	$	$		$	$
Current account	3 180		Opening balances	46 000	32 000
Realisation:			Current account		2 450
furniture & equip.	6 800				
Loss on realisation	17 400	11 600			
Bank	18 620	22 850			
	46 000	34 450		46 000	34 450

★ Exam tips

Key features of the model answer

➤ The bank balance has been omitted from the realisation account and recorded in the bank account only.

➤ The treatment of trade receivables and trade payables is designed to contrast the value in the original statement of financial position with the amounts actually received/paid.

➤ The asset taken over by Darim has been transferred to his account at transfer price.

➤ The loss on dissolution has been split in the correct ratio.

➤ The payments to the partners in the bank account are the amounts required to close their capital accounts.

➤ The current account balances have been transferred to the capital accounts.

? Exam-style questions

1   State how profits and losses should be shared in a partnership where there is no agreement according to the Partnership Act 1890.

2   Describe the purpose of including interest on drawings in a partnership agreement.

3   Amina, Kim and Mark have been in partnership for a number of years. However, Mark will be retiring in a few months' time. The partners have been discussing how to pay the amount which will be due to Mark on his retirement. They have considered admitting a new partner or taking out a bank loan at 8 per cent per annum interest. Advise the partners which course of action they should take.

4   Alex and Celia have been in partnership for a number of years sharing profits and losses equally, after charging interest on total drawings at 10 per cent per annum. The partnership's financial year ends on 31 December. On 1 October 2016 the partners agreed to change their partnership agreement with effect from that date. The terms of the new agreement were as follows:

   ● interest on capital at 8 per cent per annum to be allowed on the partners' fixed capitals of Alex $80 000, Celia $110 000

   ● a salary of $15 000 per annum to be given to Celia as a reward for managing the business

   ● interest on drawings no longer to be charged

   ● remaining profits or losses to be shared in the ratio Alex:Celia, 3:2 respectively.

   Additional information:

   ● Drawings for the period 1 January to 30 September 2016: Alex, $18 000; Celia $22 000.

   ● Drawings for the period 1 October to 31 December 2016: Alex $8 000; Celia $6 500.

   ● Profit for the year ended 31 December 2016 was $64 000; the partners agreed to assume that the profit accrued evenly throughout the year.

   ● The balances of the partners' current accounts at 1 January 2016 were: Alex $1800 (credit); Celia $640 (debit).

   Prepare the partnership's appropriation account for the year ended 31 December 2016 and Celia's current account for the year ended on that date.

5   Mike, Nadia and Ollie have been in partnership sharing profits and losses in the ratio 5:3:2 respectively. On 31 July 2016, Ollie retired and on that date the partnership's statement of affairs was as follows:

	$	$
Non-current assets at book value		185 000
Current assets		
Inventory	18 600	
Trade receivables	9 500	
Bank	4 800	
		32 900
Total assets		217 900
Capital accounts		
Mike	84 000	
Nadia	72 000	
Ollie	46 000	
		202 000
Current liabilities		
Trade payables		15 900
Total capital and liabilities		217 900

   The partners agreed that non-current assets should be revalued at $240 000; inventory should be valued at $13 790 and a provision for doubtful debts should be created of 2 per cent of trade receivables. Goodwill was valued at $70 000 and it was agreed that no goodwill account should be maintained in the partnership books. Mike and Nadia have decided to continue in partnership sharing profits and losses equally. The amount due to Ollie on his retirement will be financed by additional and equal capital contributions from Mike and Nadia, paid directly to Ollie.

   Prepare the capital accounts of the partners recording these arrangements.

6   Nizam and Rachel have been in partnership sharing profits and losses in the ratio 4:1 respectively. They have decided to admit Tara as a partner with effect from 1 November 2016. The partnership's statement of financial position immediately before Tara's admission was as follows:

	$	$
Non-current assets at book value		
Property	150000	
Equipment and fittings	27400	
		177400
Current assets		
Inventory		11200
Total assets		188600
Capital accounts		
Nizam	136200	
Rachel	48100	
		184300
Current liabilities		
Bank overdraft		4300
Total capital and liabilities		188600

Prior to Tara's admission it was agreed that property should be revalued by $24000, and inventory should be reduced in value by $1500. It was also agreed that goodwill should be valued at $80000 and that no goodwill account should be maintained in the partnership's books. Tara will provide capital of $75000 consisting of some equipment valued at $19600 and the remainder in cash. The partners have agreed to share profits and losses in future in the ratio Nizam:Rachel:Tara, 4:1:3.

Prepare the capital accounts of the partners to record these arrangements and a statement of financial position on 1 November 2016 immediately upon Tara's admission.

7 Jess and Rakesh, who share profits and losses equally, have agreed to dissolve their partnership with effect from 1 July 2016. On this date the partnership's statement of financial position was as follows:

	$	$
Non-current assets at book value		
Machinery	45300	
Motor vehicles	18600	
		63900
Current assets		
Inventory	14400	
Trade receivables	10920	
Cash at bank	8260	
		33580
Total assets		97480
Capital accounts		
Jess	49600	
Rakesh	32900	
		82500
Current accounts		
Jess	4950	
Rakesh	(770)	
		4180
Current liabilities		
Trade payables		10800
		97480

Additional information:

- Machinery was sold for $31520.
- The partners took over the motor vehicles at agreed valuations: Jess, $8200, Rakesh $7000.
- Inventory was sold for $11960.
- Trade receivables settled their accounts less discounts allowed of $610.
- The accounts of trade payables were settled less discounts received of $940.
- The costs of the dissolution totalled $1840.

Prepare (i) realisation account, (ii) partners' capital accounts and (iii) bank account, to record the dissolution of the partnership.

7 Limited companies Part 1

AS 1.4.4A, 1.4.4B SB pages 174–201

Key topics

- ➤ features of a limited company
- ➤ types of share
- ➤ reserves
- ➤ issuing shares
- ➤ debentures
- ➤ financial statements
- ➤ greatest challenges
- ➤ important techniques.

✓ What you need to know

How to prepare an income statement and statement of financial position for a limited company and be able to describe the distinction between capital and revenue reserves. Explain the different types of shares a company may issue; make bonus issues and rights issues of shares. Prepare ledger accounts to record different types of share issue. Prepare statements of changes in equity.

7.1 Features of a limited company

A limited company is an organisation owned by its shareholders. Special features of limited companies include the following.

Limited liability

Unlike sole traders and partners who have unlimited liability, shareholders have limited liability and cannot lose their personal possessions if a company fails.

Shareholders

These are owners of a limited company, and they receive a dividend as their share of the profits of the company.

Separate legal entity

This means that a company has an existence which is separate to that of its members (i.e. shareholders). A company can sue and be sued, for example.

Directors

Directors are responsible for running and managing a limited company. They are appointed by the shareholders of the company and report annually to them. Directors may also own shares in the company.

Memorandum of association

This is a formal document which must be filed with the Registrar of Companies before the company can come into existence. The memorandum details the most significant facts about the company: name, address, share capital, objectives.

> **Key term**
>
> **Limited liability:** shareholders are only responsible for the debts of the company to the limit of their investment in shares in the company.

Articles of association

This is also a formal document which must be filed with Registrar of Companies. It sets out the company's own internal rules for example: voting rights, directors' powers, how directors' meetings and the annual general meeting will be conducted, etc.

Public limited companies (plc)

A public limited company may offer their shares to members of the public. Shares can be traded on recognised stock exchanges.

Private limited companies (Ltd)

A private limited company cannot offer their shares to members of the public. They tend to be businesses where ownership is in the hands of members of a family and sometimes friends of the family.

7.2 Types of share

Ordinary shares

Are the most common type of share. Each share represents one vote, so the more shares held by a shareholder the greater the degree of control which can be exercised at the AGM (annual general meeting). Dividends paid on ordinary shares are variable and dependent on the profit made by the company and the directors' views about what can be afforded. In the event that a company goes into liquidation, ordinary shareholders are paid last.

Preference shares

Holders of preference shares received a fixed rate of dividend and take preference over ordinary shareholders in the event that a company goes into liquidation. Preference shareholders do not have voting rights.

Cumulative preference shares

Where a company cannot pay a dividend on preference shares because of the lack of distributable profits, the dividends can be carried forward and accumulate until profits are sufficient for the dividends to be financed.

Redeemable shares

Redeemable shares are shares which a company can buy back from shareholders at a future date.

7.3 Reserves

Revenue reserves

Directors do not normally distribute all the profits made by a company, in order to ensure that funds are retained to maintain the business's assets, provide for expansion, etc. The most common form of revenue reserve is 'retained profits', but some companies also record undistributed profits in a 'general reserve'. Directors may choose to distribute revenue reserves at any time in the form of dividends.

Capital reserves

The two most common types of capital reserve are: share premium (arising from issuing shares at above their nominal value); revaluation reserve (arising from increasing the value of non-current assets to better reflect their market value). A revaluation reserve is a good example of an unrealised gain (i.e. there is no increase in liquid funds) and is recorded by debiting the property account and crediting the revaluation reserve.

> **Key terms**
>
> **Revenue reserves**: undistributed profits arising from trading activities.
>
> **Capital reserves**: profits arising from non-trading activities which may be not be distributed as dividends to shareholders.

Dividends

Dividends are a shareholder's reward for investing in a company. Dividends are usually paid annually, but a mid-year (interim) dividend may also be paid. Dividends are a percentage based on the nominal value of their investment.

7.4 Issuing shares

Authorised capital

This is the maximum amount of shares that may be issued by a company according to its memorandum of association.

Issued capital

This is the amount of share capital that has actually been issued by a company.

Nominal value

This is the face value of a share (sometimes called the par value).

Share premium

This arises when shares are sold at a value above their nominal value. The share capital account records the face value of the shares; the share premium account records the amount over and above the face value.

Rights issue

This is an issue of shares for cash designed to provide a company with additional finance. In a rights issue existing shareholders are offered the opportunity to buy shares at a favourable price (i.e. at a price below the current market price). The offer is to buy shares in proportion to those already held, so, for example, a 2 for 3 rights issue, means that a shareholder could buy 2 additional shares for every 3 currently owned. A rights issue costs a company less in administration expenses than an issue to the general public. If an existing shareholder does not wish to take up the offer, the right to the shares can be passed on to another individual. If all existing shareholders take up the offer of additional shares, the control of the company remains exactly the same as before the rights issue.

Bonus issue

A bonus issue does not raise cash for a company, it is really a technical process by which a company's reserves are made more permanent by converting them into shares. In a bonus issue existing shareholders receive a number of additional shares in proportion to their existing shareholding. For example, if a shareholder owned 12 000 shares, and a bonus issue was made of 1 share for every 4 currently held, the shareholder would receive an additional 3000 shares. Bonus issues are often made where a company has large reserves (particularly capital reserves) and where to pay dividends would not be possible because of a shortage of liquid funds. In effect where there is a bonus issue issued share capital increases and reserves decrease.

7.5 Debentures

Debentures are long-term loans which can be made by a limited company. In return for the cash received, the lender receives a certificate (called a debenture) which sets out the terms of the loan: the interest rate, the date of redemption (i.e. repayment), security for the loan. Debenture interest is charged to the income statement in the same way as interest on any other loan. Debentures are shown in the statement of a financial position as a non-current liability, except in the final year when they are due for redemption when they appear as a current liability.

> **X Common error**
>
> To regard debentures as part of a company's capital structure.

7.6 Financial statements

Limited companies prepare the following end of year financial statements:

> ➤ income statement.
>
> ➤ statement of changes in equity
>
> ➤ statement of financial position.

7.7 What are the greatest challenges?

> ➤ Fully understanding the background to limited companies including the meaning of technical terms.
>
> ➤ Making accurate records of the different types of share issue.
>
> ➤ Preparing income statements set out correctly to show the key subtotals.
>
> ➤ Preparing statements of changes in equity.
>
> ➤ Preparing statements of financial position which set out the equity section correctly.

7.8 Review some important techniques

Income statement format

	$
Revenue	xx
Cost of sales	(xx)
Gross profit	xx
Other income	xx
Expenses (including auditors' fees and directors' remuneration)	
Distribution costs	(xx)
Administration expenses	(xx)
Profit from operations	xx
Finance costs (i.e. interest charges including debenture interest and preference share dividends paid)	(xx)
Profit before tax	xx
Taxation	(xx)
Profit for the year	xx

> **X Common error**
>
> To forget to record preference share dividends paid in the finance costs section of the income statement.

Statement of changes in equity format

	Share capital $	Share premium $	General reserve $	Retained earnings $	Total $
Opening balance	xx	xx	xx	xx	xx
Share issue	xx	xx			xx
Profit for year				xx	xx
Transfer to general reserve			xx	(xx)	
Dividends paid				(xx)	(xx)
Closing balance	xx	xx	xx	xx	xx

> **Exam tip**
>
> Take care to label each of the subtotals in an income statement correctly, i.e. cost of sales, gross profit, profit from operations, profit before tax, profit for the year.

> **Remember**
>
> When you prepare a statement of changes in equity do check that you have shown brackets for negative items.

Statement of financial position (equity and liabilities) format

	$
Equity and liabilities	
Equity	
Issued ordinary share capital	xx
Preference share capital	xx
Share premium	xx
Revaluation reserve	xx
General reserve	xx
Retained earnings	xx
	xx
Non-current liabilities	
7% Debentures (2025)	xx
Current liabilities	
Trade payables	xx
Taxation	xx
	xx
Total equity and liabilities	xx

> **Common error**
>
> Make sure you don't fail to provide a satisfactory description of each entry in the statement of changes in equity.

Making a bonus issue

> ▼ **Example 7.1**
>
> The directors of ABC Limited have decided to make a bonus issue of three shares for every ten ordinary shares held.
>
> | Statement of financial position **before** the bonus issue (summarised) | | | Statement of financial position **after** the bonus issue (summarised) | | |
> |---|---|---|---|---|---|
> | | | $000 | | | $000 |
> | Non-current assets | | 1960 | Non-current assets | | 1960 |
> | Current assets | | 510 | Current assets | | 510 |
> | Total assets | | 2470 | Total assets | | 2470 |
> | Equity | | | Equity | | |
> | Ordinary shares of 50c each | | 1600 | **Ordinary shares of 50c each** | | **2080** |
> | Share premium | | 220 | **Share premium** | | - |
> | General reserve | | 170 | **General reserve** | | - |
> | Retained earnings | | 480 | **Retained earnings** | | **390** |
> | Total equity | | 2470 | Total equity | | 2470 |
>
> Notes
>
> Currently the company has 3200000 issued shares (each of 50c nominal value)
>
> Notes
>
> The bonus issue will be 3/10 x 3200000 shares (of 50c each), i.e. 960000 shares, value $480000.
>
> Any reserves could be used to 'finance' the bonus issue. By using the share premium first, the directors maintain flexibility for the future in proposing the payment of dividends.
>
> The bonus issue has no effect whatever on the assets of the company (i.e. no cash is received).

The double-entry required to make the bonus issue is as follows:

Transfer funds from the reserve accounts:

Dr　Share premium $220

Dr　General reserve $170

Dr　Retained earnings $90

Cr　Bonus shares account $480

Issue bonus shares:

Dr　Bonus shares account $480

Cr　Issued share capital $480

Making a rights issue

▼ Example 7.2

The directors of RDC Limited have decided to make a rights issue and have offered ordinary shareholders the opportunity to buy one additional share for every two shares held at price of $1.40. (Assume the shareholders take up the rights issue.)

Statement of financial position **before** the rights issue (summarised)		Statement of financial position **after** the rights issue (summarised)	
	$000		$000
Non-current assets	2 120	Non-current assets	2 120
Current assets	680	**Current assets**	**2 080**
Total assets	2 800	Total assets	4 200
Equity		Equity	
Ordinary shares of $1 each	2 000	**Ordinary shares of $1 each**	**3 000**
Share premium	370	**Share premium**	**770**
General reserve	290	General reserve	290
Retained earnings	140	Retained earnings	140
Total equity	2 800	Total equity	4 200

Notes:

The rights issue provides shareholders with the opportunity to purchase in total ½ x 2 000 000 shares, i.e. 1 000 000 shares at a price of $1.40.

Although the purchase price may seem high, the market value of the shares could be higher (for example, $1.75 per share).

Notes:

A rights issue provides extra cash for the business.

The amount raised is 1 000 000 x 1.40 = $1 400 000.

Ordinary shares increase by $1 000 000 and the share premium increases by $400 000.

The double entry required to make the rights issue is as follows:

Dr Bank $1 400 000

Cr Issued share capital $1 000 000

Cr Share premium $400 000

Revision checklist

I can:

➤ describe key features of limited liability companies such as types of share, meaning of limited liability, capital reserves, revenue reserves, bonus issues, rights issues, etc. ☐

➤ prepare a limited company's income statement identifying each subtotal ☐

➤ prepare a statement of changes in equity using the correct layout and labelling entries correctly ☐

➤ prepare a statement of financial position setting out the equity section correctly ☐

➤ make entries to record a bonus issue ☐

➤ make entries to record a rights issue. ☐

⬆ Raise your grade

The following balances have been extracted from the books of H Limited at 31 December 2016.

	$000
8% Debentures (2022–2023)	300
Administration expenses	113
Cost of sales	356
Distribution costs	39
Dividends paid	
Ordinary shares	30
Preference shares	6
General reserve	127
Interest charges paid	12
Issued share capital	
Ordinary shares capital (shares of 50c each)	500
6% Preference share capital (shares of $1 each)	100
Other non-current assets	
Cost	640
provision for depreciation	104
Property at net book value	705
Retained earnings at 1 January 2016	83
Revenue	722
Share premium	110

Additional information:

- Depreciation should be provided on other non-current assets at 25 per cent per annum using the reducing balance method and allocated equally between administration expenses and distribution costs.

- Six months' interest on debentures is outstanding at 31 December 2016.

- A provision for taxation of $21 000 should be made on 31 December 2016.

- On 31 December 2016 the directors have agreed to the following:

  - To revaluation property at $960 000.

  - To make a one for one bonus issue of shares. The directors intend to maintain reserves in their most flexible form.

Prepare **(i)** an income statement for the year ended 31 December 2016; **(ii)** a statement of changes in equity for the year ended 31 December 2016; **(iii)** the equity and liabilities sections of a statement of financial position at 31 December 2016.

Student answer

H Limited

Income statement

for the year ended 31 December 2016

	$000
Revenue	722
Cost of sales	356
Gross profit	366
Administration expenses	113
Distribution costs	39
Depreciation ❶	134
Debenture interest ❷	24
❸❹	56
Taxation	21
Profit for year	35

Statement of changes in equity for the year ended 31 December 2016

	Ordinary Shares	Preference Shares	Share Premium	Revaluation Reserve	General Reserve	Retained ❺ Earnings
	$000	$000	$000	$000	$000	$000
Balance ❻	500	100	110		127	83
Revaluation				255		
Share issue	500		(110)	(181) ❼	(127)	(82)
Profit for year						35
Dividends paid						(36) ❹
Balance ❻	1 000	100	0	74	0	0

Statement of financial position at 31 December 2016 (Extract)

	$000
Equity	
Ordinary share capital	1000
6% Preference share capital	100
Revaluation reserve	74
8% Debentures (2022–2023) ⑧	300
	1474
Current liabilities ⑨	
Taxation provision	21
	1795

How to improve this answer

① Lack of workings for more complex calculations.

② Debenture interest should be recorded as part of finance costs.

③ No attempt to find subtotal for operating profit.

④ No attempt to record finance costs and preference share dividend paid omitted (it has been incorrectly included in the statement of changes in equity).

⑤ Total column has been omitted from the statement of changes in equity.

⑥ Labelling of items lacks detail.

⑦ Instruction "to maintain reserves in their most flexible form" has been ignored.

⑧ Debentures have been included in the equity section rather than as a non-current liability.

⑨ Current liability list is incomplete with interest due omitted.

Model answer

H Limited

Income statement for the year ended 31 December 2016

	$000
Revenue	722
Cost of sales	(356)
Gross profit	366
Administration expenses (113 + 67)	(180)
Distribution costs (39 + 67)	(106)
Operating profit	80
Finance costs (**W2**)	(30)
Profit before tax	50
Taxation	(21)
Profit for year	29

W1 Depreciation

25% x (640 – 106), i.e. 134

W2 Finance costs

Interest charges paid 12 + debenture interest due 12 + preference share dividend 6

Statement of changes in equity for the year ended 31 December 2016

	Ordinary Shares	Preference Shares	Share Premium	Revaluation Reserve	General Reserve	Retained Earnings	Total
	$000	$000	$000	$000	$000	$000	$000
Balance 1 January	500	100	110		127	83	920
Revaluation of property				255			255
Bonus issue	500		(110)	(255)	(127)	(8)	0
Profit for year						29	29
Dividends paid						(30)	(30)
Balance 31 December	1 000	100	0	0	0	74	1 174

Statement of financial position at 31 December 2016 (Extract)

	$000	$000
Equity		
Ordinary share capital		1 000
6% Preference share capital		100
Retained earnings		74
		1 174
Non-current liabilities		
8% Debentures (2022–2023)		300
Current liabilities		
Taxation provision	21	
Other payable (interest due)	12	
		33
Total equity and liabilities		1 507

★ Exam tips

Key features of the model answer

➤ Workings have been included for more complex calculations (depreciation, finance costs).

➤ Operating costs and finance costs have been clearly identified.

➤ The preference share dividend has been included in the finance costs.

➤ A total column has been included in the statement of changes in equity and the items have been more fully detailed (for example, opening and closing balances are given dates).

➤ The bonus issue has been financed from capital reserves and then from revenue reserves to meet the 'reserves in flexible form' requirement.

➤ The liabilities section of the statement of financial position is now correctly shown to include debentures (non-current liability) and other payables (current liability).

 Link

Statements of cash flow are covered in Unit 13.

❓ Exam-style questions

1   State what is meant by the term limited liability in relation to a limited liability company.

2   Describe the key features of a capital reserve.

3   Explain how shareholders might benefit from a rights issue of shares.

4   The following trial balance was extracted from the books of Q Limited on 31 December 2016.

	$000	$000
6% Debentures (2024–2025)		300
Administration expenses	49	
Cash and cash equivalents	36	
Debenture interest paid	9	
Distribution costs	81	
Dividends paid	45	
Fittings and equipment		
Cost	520	
Provision for depreciation		127
Inventory 1 January 2016	62	
Issued capital (ordinary shares of 50c each)		700
Property		
Cost	950	
Provision for depreciation		85
Purchases	845	
Retained earnings		136
Revenue		1148
Share premium		112
Trade payables		31
Trade receivables	42	
	2639	2639

Additional information:

• On 15 November the directors had made a rights issue of shares of 3 shares for every 7 shares currently held for $1.60 per share. The rights issue was fully subscribed but no entries have yet been made in the books.

• On 31 December 2016 the directors agreed to revalue the property at $1 300 000.

• Auditor's fees of $11 000 were due but not yet paid on 31 December 2016. These should be charged to administration expenses.

• On 31 December 2016 inventory was valued at $71 000.

• Depreciation should be provided on fittings and equipment at 20 per cent per annum using the straight-line method. Depreciation is charged to administration expenses.

• Debenture interest for the period 1 July to 31 December 2016 is due but unpaid.

• The taxation charge for the year is $13 000.

Prepare:

a) An income statement for the year ended 31 December 2016.

b) A statement of changes in equity for the year ended 31 December 2016.

c) A statement of financial position at 31 December 2016.

8 Analysis and communication of accounting information to stakeholders

Key topics

- ➤ ratio analysis
- ➤ stakeholders
- ➤ accounting concepts and financial statements
- ➤ greatest challenges
- ➤ important techniques.

✓ **What you need to know**

There are different groups interested in the financial performance of a business and different ways of analysing the performance. You will need to be able to evaluate the profitability, liquidity and efficiency of a business using ratios whilst realising that this process has its limitations.

8.1 Ratio analysis

Use of ratios

Accounting ratios (ratio analysis) are one of the main means of analysing a business's performance because they provide an opportunity to compare:

- ➤ results from one year with another year
- ➤ a business's performance with that of other businesses
- ➤ a business's performance with the industry averages.

They provide useful information for stakeholders.

Ratios are used to measure three key aspects of a business's performance:

- ➤ profitability
- ➤ liquidity
- ➤ efficiency.

Key term

Ratio analysis: a technique for measuring the financial results of a business in order to assess its performance.

Different types of ratios

Table 8.1 lists the key accounting ratios.

▼ **Table 8.1** Accounting ratios

Ratio	Formula	What does it tell you?
Profitability		
Gross profit margin	$\dfrac{\text{Gross profit}}{\text{Revenue}} \times 100$	Gross profit in cents made on each $1 of revenue (pricing policy of business).
Mark-up	$\dfrac{\text{Gross profit}}{\text{Cost of sales}} \times 100$	Gross profit in cents made on each $1 of cost of sales (pricing policy of business).

Profit margin	$\dfrac{\text{Profit for year}}{\text{Revenue}} \times 100$	Profit in cents made on each $1 of revenue.
Expenses in relation to revenue	$\dfrac{\text{Expense}}{\text{Revenue}} \times 100$	Expense in cents incurred for each $1 of revenue.
Return on capital employed	$\dfrac{\text{Profit year before interest}}{\text{Capital employed*}} \times 100$	The profit in cents made on each $1 of capital used within the business.

*capital employed for a limited company is: equity (issued shares + reserves) + long-term liabilities.

	Liquidity	
Current ratio	Current assets:current liabilities	The amount of current assets (in dollars) in relation to current liabilities. How easily can the business meet its commitments* in the short to medium term.
Liquid ratio	Current assets (excluding inventory): current liabilities	The amount of liquid assets in relation to current liabilities. How easily can the business meet its commitments* in the short term.

*'Commitments' include all the current liabilities which will fall due for payment in the short/medium term plus all the day to day running costs for which payment will be required plus owner's drawings (or dividend payments due) plus any planned capital expenditure, etc.

	Efficiency	
Rate of inventory turnover (times)	$\dfrac{\text{Cost of sales}}{\text{Average inventory}}$	How frequently the average inventory has been sold during a time period.
Rate of inventory turnover (days)	$\dfrac{\text{Average inventory}}{\text{Cost of sales}} \times 365$	The number of days it takes for the average inventory to be sold during a time period.
Trade payables turnover	$\dfrac{\text{Trade payables}}{\text{Credit purchases}} \times 365$	Average time it takes to settle accounts of credit suppliers.
Trade receivables turnover	$\dfrac{\text{Trade receivables}}{\text{Credit sales}} \times 365$	Average time it takes credit customers to settle their accounts.
Non-current asset turnover	$\dfrac{\text{Net revenue}}{\text{Non-current assets net book value}}$	Revenue earned in dollars for every $1 of non-current assets.

Interpreting ratios

There are a number of ways in which you can comment on ratios, as shown in Table 8.2 below.

▼ **Table 8.2** Interpreting ratios

Type of comment	Example
Describing what a ratio means.	Mark-up gives the amount of gross profit in cents per $1 of revenue.
Describing a change in a ratio.	Profit margin has increased by 5% since last year.
Analysing a change in a ratio including providing an explanation for the change.	Gross profit margin has decreased by 7% since last year and this may have been caused by lowering selling prices in the hope of increasing demand.
Evaluating a change in a ratio by clarifying whether the change is a strength or weakness.	Trade receivable turnover has declined since year and this is a weakness in the business performance since it has had a negative impact on the business's liquidity and because it could lead to an increase in irrecoverable debts.
Making recommendations to overcome weaknesses and pointing out any possible adverse consequences of the recommendation.	The rate of inventory turnover could be improved by reducing the average inventory held by the business which will have the additional benefit of reducing storage costs. However, reducing the average inventory could reduce customer choice, leading to a loss of revenue.

Limitations of ratios

These include:

➤ Financial statements may be out of date by the time they are analysed. Important events affecting the business may have taken place since they were produced (good or bad economic news, for example) which could affect performance.

➤ Financial statements only include data which has a monetary value (the money measurement concept), so important features of a business cannot be directly reflected in the accounts (e.g. morale of workforce, management expertise).

➤ There could be key financial and policy differences between businesses. For example, one business owns its premises so it has a high capital employed, but another business rents its premises, so it has a lower profit; one business uses straight-line depreciation, another business uses the reducing balance method.

➤ Inflation is usually ignored when preparing financial statements, but this can distort comparisons (e.g. payments for wages may seem much smaller three or four years ago compared to now, but high inflation in the intervening years may be an important factor and not just pay rises).

8.2 Stakeholders

Stakeholders, as well as being interested in the performance of a business, also have an impact on the performance of the business.

▼ **Table 8.3** Stakeholder questions

Stakeholder	Questions stakeholder ask
Owners (sole trader, partner)	Has my investment in the business been worthwhile? Should I invest more? What do I need to do to overcome weaknesses in performance?

Owners (shareholders)	Is my investment doing well?
	Should I invest more or invest less?
Manager/director	What can I learn from the analysis for future planning?
	How can I further develop the strengths in performance?
	How can I overcome weaknesses?
Employees	Is my job secure?
	Is the performance good enough to provide pay increases, better working conditions, more training?
Lenders (including bank manager)	Can the business afford to make regular loan repayments and interest payments?
Suppliers	Can the business afford to settle amounts owing within credit period?
Customers	Is the business likely to continue trading?
Potential investors	Which business is likely to provide the best return on any investment?
Government	How much tax is due?
	What contribution does the business make to the nation's economy?
Environmental groups	Does the business have a good reputation in regard to avoiding damage to the local environment?
Local community	Does the business support the local economy and offer employment for local people?

8.3 Accounting concepts and the analysis of accounting information

All users of accounting information rely on the fundamental principle that accounting statements should provide a true and fair picture of the business's financial position. Users will also expect the consistency concept to have been applied so that when making comparisons of a business's performance over a number of years they can be confident that the same policies have been applied, ensuring that their assessments have validity. The money measurement concept means that only information with a financial value can be reported (see limitations above).

8.4 What are the greatest challenges?

➤ Remembering the formula for each of the ratios.

➤ Interpreting changes in ratios correctly.

➤ Writing effective reports on a business's performance including making recommendations for how a business's performance could be improved.

> ★ **Exam tip**
>
> It is advisable to provide detailed information when calculating ratios to include:
>
> ➤ the formula
>
> ➤ the data selected from the information.

8.5 Review some important techniques

Setting out answers to ratio calculations

▼ **Example 8.1** Making ratio calculations

Assume the question gives a set of financial statements for a business and asks for the following ratios to be calculated: rate of inventory turnover; profit margin; trade receivables turnover.

The three ratios are set out effectively in Table 8.4:

▼ **Table 8.4** Ratio calculations

Rate of inventory turnover	$\dfrac{\text{Cost of sales}}{\text{Average turnover}}$	$\dfrac{\$140\,000}{\$28\,000}$	5 times
Profit margin	$\dfrac{\text{Profit for the year}}{\text{Revenue}} \times 100$	$\dfrac{\$19\,400}{\$215\,000}$	9.02%
Trade receivable turnover	$\dfrac{\text{Trade receivables}}{\text{Credit sales}} \times 365$	$\dfrac{\$12\,190}{\$168\,530}$	27 days*
	*The exact result is 26.40, but always round up.		

Remember

The ratio answer should be correct to the number of decimal places required (but note that trade payables turnover and trade receivables turnover are always rounded up to the next whole day).

✗ Common error

Don't forget to omit the '%' sign (if appropriate) or other descriptor.

Reviewing business performance

When commenting on a business performance it is usual to consider the following issues.

| Profitability | Is the business increasing in value over time through successful trading and providing a large enough increase in wealth to reward the owners? |
| Liquidity | Are the business's liquid resources well managed so that running costs are covered, liabilities can be met when due and drawings (or dividends) can be paid at a satisfactory level? Is the business avoiding unnecessary borrowing, but also not tying up liquid resources unnecessarily? |

★ Exam tip

When interpreting ratios it is important to go further than just describing what the ratios shows. Always consider whether any change in a ratio is a strength or a weakness for the business, and (if the question requires) suggest ways in which performance could be improved.

▼ **Example 8.2** Interpreting profitability ratios

The following information is available about a business's performance over the last two years and the industry average for that business sector.

▼ **Table 8.5** Comparing business performance

Ratio	For year ended 31 December 2015	For year ended 31 December 2016	Industry average
Gross profit margin	28%	32%	29%
Profit margin	13%	11%	14%
Return on capital employed	11%	14%	12%

In the example, the gross profit margin has increased by 4 per cent and has moved from being below the industry average to being above by 3 per cent. This is a potential strength for the business as it means it will being making an additional 4 cents profit for every $1 of revenue. However, this change in pricing policy may make the business more uncompetitive and reduce total revenue.

The profit margin has fallen by 2 per cent compared to the previous year and has fallen further below the industry average. This means that 2 cents less profit is being made for every $1 of revenue, and this is happening despite the improvement in the gross profit margin. This is a weakness for the business and could mean that there has been less efficient control of expenses since last year. It is recommended that expenses are investigated to see why they may have increased with a view to discovering unnecessary expenditure and eliminating it in the future.

The return on capital employed has improved by 3 per cent since last year and is now higher that the industry average by 2 per cent instead of being lower by 1 per cent. The business is now earning an extra 3 cents per $1 of capital employed, despite the decline in the profit margin. This is a strength for the business. It is likely that the improvement is due to its assets being used more efficiently – perhaps underused assets have been sold off.

Notes:

➤ The comment on each ratio goes further than just describing the change. (This is the most basic form of response and on its own would normally be considered inadequate.)

➤ By suggesting why the change may have happened gives a chance to demonstrate understanding of the ratio.

➤ It is important to evaluate the change in each ratio by making it clear whether the business is improving (a strength) or declining (a weakness).

➤ Where the performance has weakened it is often a requirement to recommend what measures should be taken to improve performance.

Revision checklist

I can:

➤ recall and use the ratios and their formulas for profitability ☐

➤ recall and use the ratios and their formulas for liquidity ☐

➤ recall and use the ratios and their formulas for efficiency ☐

➤ state the ratio in the correct form ☐

➤ describe what the ratio tells a stakeholder about performance ☐

➤ analyse changes in a ratio and suggest possible reasons for the changes ☐

➤ evaluate a change in a ratio (i.e. whether a strength or a weakness) ☐

➤ recommend how any weaknesses in performance could be overcome. ☐

A sole trader has provided the following information concerning the business's financial position at 31 December 2016:

	$	$
Current assets		
Inventory	14 600	
Trade receivables	12 250	
Other receivables	710	
Cash at bank	1 180	
		28 740
Current liabilities		
Trade payables	15 360	
Other payables	1 290	
		16 650

Additional information:

- Credit sales during the year ended 31 December 2016 totalled $163 300.

- The following table shows ratios which were calculated at 31 December 2015 and the industry averages for this sector.

	At 31 December 2015	Industry average
Current ratio	1.4 : 1	1.5 : 1
Trade receivable turnover	35 days	31 days

Assess the performance of the business for the year ended 31 December 2016 and suggest possible reasons for changes.

Student answer

Current ratio at 31 December 2016 is:

28 740 : 16 650 which is 1.7 : 1 ❶

The ratio has improved and is now better than the industry average. The business is in a better position to meet its debts. ❷ Maybe the business has made more profits this year. ❸

Trade receivables ratio at 31 December 2016 is 27.38 ❶ ❹

This is good news for the business as it is receiving cash back far more quickly than last year and more quickly than the industrial average. ❺ Maybe the business has offered cash discounts. ❸

How to improve this answer

❶ The formula used for each ratio has not been included as part of the answer and the current ratio answer has been shown to just one decimal place.

❷ The statement about the business's better position is very brief and could be developed much further.

③ The reason suggested for the change in each of the ratios is very brief; much more could have been written.

④ The word 'days' is missing from the answer to the trade receivables ratio; the answer for this ratio should normally be rounded up to the next whole day.

⑤ No attempt has been made to suggest a possible weakness arising from the change in the trade receivables ratio.

Model answer

Current ratio formula:

Current assets:current liabilities

i.e. 28 740 : 16 650 which is 1.73 : 1.

The ratio has increased since last year and has moved from below the industry average to above the industry average. This will mean that the business will find it easier to pay its debts and running costs when they fall due, which is a strength. However, because the ratio is now higher than the industry average it could also be a weakness if the business is holding resources as current assets unnecessarily. Maybe some of the resources could be made better use of. The change may have occurred because the business has increased its profits and/or the owner may have reduced drawings, arranged a long-term loan or sold off unwanted non-current assets. All of these would have increased liquid resources.

Trade receivables formula:

$$\frac{\text{Trade receivables}}{\text{Credit sales}} \times 365 \text{ i.e. } \frac{12\,250}{163\,300} \times 365 = 28 \text{ days}$$

The ratio has improved since last, year since credit customers are now paying much more quickly than before (by 7 days) and more quickly than the industry average (by 3 days). This is a possible strength for the business as it will receive cash more quickly, enabling the business to discharge its liabilities and pay its running costs more easily. However, it may also prove to be a weakness, if credit control is now too tight and potential customers are deterred from buying from the business leading to a fall in revenue. The improvement may have arisen from offering cash discounts, but if so this will reduce profits. It could also have been caused by reducing the credit period allowed to customers.

★ Exam tips

Key features of the model answer

➤ In each case full details of the ratio are given: formula, data selected.

➤ Both the possible strength and the possible weakness are explained for the change in each ratio.

➤ The possible reasons for the change in each ratio are explained more fully.

➤ In the case of the trade receivables ratio, the answer has been rounded up to the next whole day and labelled 'days'.

1   Identify two internal stakeholders in a business and state their interest in the business's financial position.

2   Describe two benefits of using ratios to assess business performance.

3   Explain what the rate of inventory turnover can tell you about a business's performance.

4   The financial statements of K Ltd for the year ended 31 December 2016 were as follows:

Income statement for the year ended 31 December 2016

	$
Revenue	824 000
Cost of sales	(372 650)
Gross profit	451 350
Operating expenses	(178 650)
Operating profit	272 700
Finance cost	(25 000)
Profit before tax	247 700
Taxation	(86 000)
Profit for the year	161 700

Additional information:

Dividends paid during 2016 totalled $130 000

Retained earnings at 1 January 2016: $124 000

Statement of financial position at 31 December 2016

	$	$
Non-current assets		1 115 800
Current assets		
Inventory	67 600	
Trade receivables	31 200	
Other receivables	2 400	
Cash and cash equivalents	26 400	
		127 600
Total assets		1 243 400

Equity		
Ordinary share capital ($1 shares)	640 000	
Share premium	88 000	
Retained earnings	155 700	
		883 700
Non-current liabilities		
10% Debenture (2026–2027)		250 000
Current liabilities		
Trade payables	23 700	
Taxation due	86 000	
		109 700
Total equity and liabilities		1 243 400

Ratios calculated at 31 December 2015 were as follows

Gross profit margin	56.13%
Operating expenses/revenue	18.52%
Rate of inventory turnover	7 times
Profit margin	16.45%
Acid test ratio	0.48 : 1
Non-current assets turnover	0.71 times
Return on capital employed	20.24%

(a) Calculate the following ratios for the current year:

- Gross profit margin
- Operating expenses/revenue
- Rate of inventory turnover
- Profit margin
- Acid test ratio
- Non-current assets turnover
- Return on capital employed

(b) Assess the performance of the company based on these ratios, indicating areas of improvement and areas of weakness. Make recommendations as to how any weaknesses could be improved.

Key topics

➤ purpose of cost accounting

➤ types of cost

➤ inventory valuation

➤ absorption costing

➤ greatest challenges

➤ important techniques.

✓ **What you need to know**

To identify and calculate different types of costs, such as material and labour costs, and understand the application of absorption costing. Using the FIFO and AVCO methods you can calculate the value of closing inventories.

9.1 Purpose of cost accounting

Cost accounting seeks to ensure that those responsible for managing a business are aware of the costs which are involved in producing, purchasing and selling products, so that they can make correct decisions to ensure the success of the business. Cost accounting is vital to the successful management of many organisations including:

➤ all types of business: manufacturing, wholesaling, retailing, service providers, etc.

➤ publicly owned organisations: hospitals, police authorities, schools, colleges, universities, etc.

9.2 Types of cost

Costs can be classified in a variety of ways:

➤ **Element** – made up of:

   ➤ direct materials, direct labour, direct overheads

   ➤ indirect materials, indirect labour, indirect overheads.

This classification can be helpful in identifying the most significant costs where efforts might best be directed at making economies.

➤ **Function** – made up of categories such as production, sales, administration, distribution, financing in a manufacturing organisation or other suitable categories for other types of business. This classification can be helpful in identifying the most expensive departments where efforts might best be directed at making economies.

➤ **Behaviour**, i.e. what causes costs to change:

   ➤ Fixed costs: are those costs which occur whatever the level of business activity, i.e. remain the same whatever the level of factory production – factory manager's salary, machinery depreciation, factory rent are good examples.

Key term

Fixed costs: costs which do not change with the level of business activity.

AS Level

- **Variable costs**: those costs which change in direct proportion to changes in the level of business activity – direct labour, direct materials are both good examples.

- **Semi-variable costs**: those costs which contain both a fixed element and a variable element – an example would be power charges, where there is a standing charge payable whatever the amount of use and a variable element of $x per unit of power used.

- **Stepped costs**: are costs which are fixed but change when a certain level of production is reached; at this higher level costs again remain fixed until a second higher level of production is reached, and so on.

Material costs

Materials costs can be direct (where the cost of materials used in one product can be identified) or indirect (where the cost cannot be identified with an individual product or are a very minor part of the finished product). Indirect materials are treated as overhead costs and examples include the glue, nails, screws, etc. used in making furniture products, or the very small quantities of spice used in producing ready-made meals.

Labour costs

Direct labour costs arise when the work done can be identified in each product (usually this means operating machines or assembling a product). Indirect labour costs, however, cannot be specifically identified in an individual product and include the remuneration of those supervising workers and machinery, maintaining machinery and equipment. Employee remuneration can be calculated by using:

- **Time rates**: where an hourly rate is applied (with a higher rate paid for overtime working).

- **Piece rates**: where payment is based on the number of items produced.

- **Bonuses**: where additional payments are made if targets are exceeded.

9.3 Inventory valuation

The fundamental rule for valuing inventory is that it should be valued at cost or net realisable value whichever is lower. However, a problem arises because the cost of inventory can vary over a period of time due to price inflation or deflation, leading to the application of one of the following methods to establish cost:

FIFO (first in first out)

FIFO is where it is assumed that the first items received were the first items to be sold and so it is assumed a closing inventory will be made up of more recent purchases. This method has the benefit of basing the valuation on most recent prices.

LIFO (last in first out)

LIFO is where it is assumed that the last items received were the first to be sold, and so it is assumed that a closing inventory tends to be made up of the earliest purchases.

> **Key term**
>
> **Variable costs**: costs which change in direct proportion to levels of business activity.

> **Key term**
>
> **FIFO**: a method of valuing inventory where it is assumed the first items received were the first items to be sold.

> **Key term**
>
> **LIFO**: a method of valuing inventory where it is assumed the last items received were the first items to be sold.

> **X Common error**
>
> In the case of FIFO and LIFO to believe that goods are actually issued in this sequence, forgetting that it is just an assumption that this has happened.

AVCO (average cost)

Where the average cost of inventory is recalculated whenever a new purchase of goods takes place. The AVCO method does require more complex calculations but it does modify the effect of fluctuations in prices and ensures that, unlike the other methods, identical products are given the same value.

Inventory valuation methods and the effect on profits

In a period of inflation FIFO will give a higher value of closing inventory than AVCO and as a result FIFO will mean that the profit for the period is also higher than the profit calculated where AVCO is used. This is because FIFO valuations are based on the most recent purchases which will be the highest prices in a period of inflation. (The reverse applies in a period of deflation.) However, it is important to remember that over several accounting periods the effect on profit of using either of these methods is evened out. This is because one period's closing inventory will become the next period's opening inventory, and whatever the effect on profit of the closing inventory in the first period will countered by the opposite effect on profit in the next period.

9.4 Absorption costing

Is a technique used to establish the total cost of producing one unit, by taking account of the indirect costs and using a procedure to:

➤ **Allocate** indirect costs to cost centres where it is clear that whole items of expenditure are derived from that the cost centre.

➤ **Apportion** indirect costs to cost centres using some rational basis (e.g. sharing machinery depreciation on the basis of the cost of machinery in each department).

➤ **Reapportion** the cost of service departments to other departments on a rational basis (e.g. sharing the cost of a personnel department on the basis of the number of employees in each department).

➤ **Calculate overhead absorption rates:** for each production department the total indirect costs are divided by either the number of labour hours or machine hours available (whichever is the more dominant feature) to provide an OAR (overhead absorption rate) to be used in calculating the cost of making a product.

Because absorption costing enables the total cost of a product to be calculated, it is useful in setting prices.

9.5 What are the greatest challenges?

➤ Understanding the different types of cost and cost behaviours.

➤ Absorption costing: using the most appropriate means of apportioning costs and calculating overhead absorption rates.

➤ Absorption costing: understanding under- and over-absorption of overheads.

9.6 Review some important techniques

Using AVCO

Valuing inventories using AVCO leads to more complex calculations than with other methods. After each new purchase the average cost of inventory has to be recalculated.

> **Key term**
>
> **AVCO:** a method of valuing inventory where the weighted average of the cost of inventory is recalculated after every purchase.

> **Key terms**
>
> **Indirect costs:** a cost that can only be attributed to *total production* and not to a single unit of production.
>
> **Cost centre:** a production or service location.

> **X Common error**
>
> Make sure you don't confuse the terms 'allocate' and 'apportion'.

> **Key terms**
>
> **Overhead absorption rate:** the rate used to absorb overheads into the cost of making one unit.
>
> **Absorption costing:** a system of absorbing factory overheads (both fixed and variable) into the total production cost of a unit.

▼ **Example 9.1** Calculating inventory values using AVCO

Table 9.1 sets out details of purchases and sales of a product and the valuation of the inventory using AVCO.

▼ **Table 9.1** Valuation of inventory using AVCO

Date	Received			Issued	Average Cost	Inventory valuation		Average cost calculation details
	Qty	Price	Value	Qty		Qty	Value	
		$	$		$		$	
Jan	30	10	300		10	30	300	
Feb	50	12	600		11.25	80	900	900/80 = $11.25
Mar				40	11.25	40	450	
June	60	15	900		13.50	100	1350	1350/100 = $13.50
July				50	13.50	50	675	
Aug	70	16	1120		14.9583	120	1795	1795/120 = $14.9583
Sept				30	14.9583	90	1346.25	

Absorption costing

Table 9.2 sets out an example of the allocation, apportionment and reapportionment of indirect cost typical of absorption costing.

▼ **Example 9.2** Absorption costing – allocate, apportionment and reapportion overheads (assume the overheads are for the year ended 31 December 2016)

▼ **Table 9.2** Absorption costing

	Production depts		Service depts		Notes
	A	B	M	N	
	$	$	$	$	
Allocation of quality controller's salary.		18000			Quality control only occurs in department B, so the entire cost can be *allocated* to this department.
Apportionment of rent based on floor area of each department.	16000	10000	2000	2000	Where costs cannot be directly allocated they must be *apportioned* using a rational basis to charge overheads to departments.
Apportionment of depreciation based on cost or carrying amount of assets in each department.	9000	8000	3000	4000	
	25000	36000	5000	6000	Subtotal of overheads before reapportionment of service department overheads.

Reapportionment Service department N	3 000	2 000	1 000	(6 000)	Reapportionment Stage 1: reapportionment of Service department N.
	28 000	**38 000**	**6 000**	**0**	New subtotal
Reapportionment Service department M	4 000	2 000	(6 000)		Reapportionment Stage 2: Reapportionment of remaining service department M.
TOTAL	**32 000**	**40 000**	**0**	**0**	Total overheads allocated and apportioned to production departments.

Calculating overhead absorption rates

Once total overheads for each production department have been calculated (see Example 9.2 on the previous page) it is possible to calculate overhead absorption rates. Rates are based on the dominant factor within a production department: i.e. labour hours where a department is labour intensive, or machine hours where a department is machine intensive.

These absorption rates are a means of attributing some portion of a department's total overheads to a unit, depending on how many labour hours/machine hours the unit takes to produce in a department.

> ★ **Exam tip**
>
> The first service department to choose when reapportioning overheads will be a department which also provides a service for the production departments and any other service department.

▼ **Example 9.3** Calculating overhead absorption rates

Using the data in Example 9.2 on the previous page and the following information:

Depart	Labour hours	Machine hours
A	64 000	33 000
B	10 000	16 000

The overhead absorption rate (OAR) calculations are as follows:

Depart	Basis	Calculation details	OAR
A	Labour hours	Total overheads/64 000 hrs i.e. $\dfrac{\$32\,000}{64\,000}$	$0.50 per labour hour
B	Machine hours	Total overheads/16 000 hrs i.e. $\dfrac{\$40\,000}{16\,000}$	$2.50 per machine hour

> ✗ **Common error**
>
> It is a common mistake to fail to label the OAR correctly. It is important to describe an OAR as $x per labour hour or $x per machine hour.

Calculating the cost of a unit

In order to calculate the cost of a unit it is necessary to take account of direct costs (materials and labour) and to add in overheads which are dependent on how long a unit takes in each production department.

> **Key term**
>
> **Direct cost**: a cost that can be attributed to a *single* unit of production.

▼ Example 9.4

Using the data from Example 9.3 on the previous page and the following information.

A unit requires 2 kg of materials at $4.80 per kg. A unit requires the following time in each production department:

Depart	Labour hours	Machine hours
A	4	3
B	7	2

Labour is paid $8 per hour in both departments.

The full cost of one unit is as follows:

	$	Notes
Materials: 2kg x $4.80 per kg	9.60	
Labour: 11 hours at $8 per hour	88.00	
Department A overheads: 4 x $0.50 per labour hour	2.00	In this labour-intensive department, machine hours used can be ignored for calculating overheads.
Department B overheads: 2 x $2.50 per machine hour	5.00	In this machine-intensive department, labour hours used can be ignored for calculating overheads.
Full cost	104.60	

Under-absorption and over-absorption of overheads

Overhead absorption rates are calculate on forecasts of the total overheads for production departments and on planned levels of production.

Where actual total overheads and actual production are exactly as forecast all overheads will be absorbed. However, the following may occur:

Total overheads are more than forecast.	Overheads are under-absorbed.
Total overheads are less than forecast.	Overheads are over-absorbed.
Actual production exceeds forecast production.	Overheads are over-absorbed.
Actual production is less than forecast production.	Overheads are under-absorbed.

Where overheads are under-absorbed an adjustment is made in the income statement (i.e. the amount is debited).

Where overheads are over-absorbed an adjustment is made in the income statement (i.e. the amount is credited to the income statement).

▼ **Example 9.5** Calculating under-absorption and over-absorption

Information for Department A
(where overheads are absorbed on the basis of labour hours)

	Budgeted	Actual
Production units	40 000	36 000
Direct labour hours	20 000	19 000
Overhead costs	$160 000	$156 000

The overhead absorption rate is based on budgeted information and is $8 per labour hour.

Calculation of **under**-absorption of overheads:

	$
Overheads actual recovered (actual labour hours x OAR per labour hour) I.e. 19 000 x $8	152 000
Actual overheads	156 000
Under-absorption of overheads	4 000

Information for Department B
(where overheads are absorbed on the basis of machine hours)

	Budgeted	Actual
Production units	6 000	6 400
Direct machine hours	11 000	12 200
Overhead costs	$132 000	$129 000

The overhead absorption rate is based on budgeted information and is $12 per machine hour.

	$
Overheads actual recovered (actual machine hours x OAR per machine hour) I.e. 12 200 x $12	146 400
Actual overheads	129 000
Over-absorption of overheads	17 400

Budgeted absorption costs are used to find the total cost of producing a unit which is then used to set a selling price.

Where absorption costs are under-absorbed the following could result:

➤ insufficient overhead is charged to the product

➤ the selling price charged is too low

➤ profits are reduced OR the lower price increases demand for the product leading to higher profits.

Where absorption costs are over-absorbed the following could result:

➤ too much overhead is charged to the product

➤ the selling price charged is too high

➤ profits are higher OR the higher price reduces demand for the product leading to less profit.

⬆ Raise your grade

K Ltd is a manufacturing company which is organised into two production departments: machining, assembly; and two service departments: maintenance and canteen. The following overheads have been forecast for the year ended 31 December 2017.

Indirect costs already allocated:

	Machining	Assembly	Maintenance	Canteen
	$	$	$	$
Indirect costs already allocated	22 400	18 900	7 985	4 200

Indirect costs to be apportioned:

	$
Depreciation of machinery	52 000
Insurance of machinery	13 500
Rent	44 100
Power	15 600

The following cost centre information is available:

	Production departments		Service departments	
	Machining	Assembly	Maintenance	Canteen
Floor space (m²)	20 000	6 000	2 000	2 000
Machinery (cost)($)	52 000	34 000	6 300	11 700
Machinery (replacement value) ($)	80 000	45 000	10 000	15 000
Number of machines	26	8	4	2
Kilowatt hours	16 000	6 000	500	1 500
Number of employees	52	26	3	5

The following budgeted information is available for October 2016:

	Machining	Assembly	Canteen
Direct labour hours	1 200	2 700	
Direct machine hours	5 400	800	
Maintenance hours	190	80	30

Canteen facilities are provided for production department employees only.

Prepare a statement showing the apportionment of overheads for 2017 and calculate the overhead absorption rates.

Student answer

❶

	Basis	Production Depts		Service Depts	
		Machining $	Assembly $	Maintenance $	Canteen $
Allocated indirect costs		22 400	18 900	7 985	4 200
Machinery depreciation	Replacemnt ❷	27 733	15 600	3 467	5 200
Insurance of machinery	Cost ❷	6 750	4 413	818	1 519
Rent	❶	29 400	8 820	2 940	2 940
Power	❶	10 400	3 900	325	975
	❸	96 683	51 633	15 534	14 834
Reapportionment:					
Canteen	No. of staff	9 889	4 945		(14 834)
		106 573	56 578	15 534	0
Maintenance	No. of machines ❹	11 879	3 655	(15 534)	
		118 452	60 233	0	

Overhead absorption rates

Machining department

Total overheads	$118 452	i.e. $99 ❸ ❻
Labour hours ❺	1 200	

Assembly department

Total overheads	$60 233	i.e. $75 ❸ ❻
Machine hours ❺	800	

How to improve this answer

❶ Some details are omitted: overall heading; bases of apportionment.

❷ The wrong basis for apportionment has been chosen for machinery depreciation and for insurance of machinery.

③ Figures have been rounded leading to inaccuracies in the final results.

④ The service departments have been reapportioned in the wrong order.

⑤ The wrong bases for determining the OARs have been used.

⑥ Each OAR lacks a description i.e. per labour hour, per machine hour.

Model answer

Apportionment of budgeted overheads for year ending 31 December 2017

| | | Production Depts | | Service Depts | |
	Basis	Machining	Assembly	Maintenance	Canteen
		$	$	$	$
Allocated indirect costs		22 400	18 900	7 985	4 200
Machinery depreciation	Cost	26 000	17 000	3 150	5 850
Insurance of machinery	Replacement	7 200	4 050	900	1 350
Rent	Floor space	29 400	8 820	2 940	2 940
Power	Kilowatt hrs	10 400	3 900	325	975
		95 400	52 670	15 300	15 315
Reapportionment:					
Maintenance	Mntnc hrs	9 690	4 080	(15 300)	1 530
		105 090	56 750	0	16 845
Canteen	No of staff	11 230	5 615		(16 845)
		116 320	62 365		0

Overhead absorption rates

Machining department

Total overheads	$116 320	i.e. $21.54 per machine hour
Machine hours	5 400	

Assembly department

Total overheads	$62 365	i.e. $23.10 per labour hour
Labour hours	2 700	

❓ Exam-style questions

1 Define the terms direct costs and indirect costs.

2 Explain what is meant by an overhead absorption rate.

3 The FIFO method of inventory valuation has been used for some years by a business. A director, has now suggested that the company switch to using the AVCO method of inventory valuation. Discuss whether changing the method of inventory valuation would be a good idea.

4 W Ltd is a manufacturing company. It operates with three production departments and two service departments. Budgeted information for December 2016 is available as follows.

Indirect costs allocated to departments:

	Production departments			Service departments	
	Cutting	Machining	Finishing	Stores	Canteen
	$	$	$	$	$
Allocated indirect costs	27 000	32 000	18 500	6 353	6 420

Indirect costs to be apportioned:

	$
Depreciation of machinery	33 600
Wages of production departments supervisors	42 000
Power	17 600

Additional information:

	Cutting	Machining	Finishing	Stores
Direct labour hours	7 600	6 700	8 300	
Direct machine hours	8 100	12 300	2 400	
Stores requisitions	4 800	1 200	900	
Number of employees	13	11	6	4

	Cutting	Machining	Finishing	Stores	Canteen
	$	$	$	$	$
Machinery cost ($)	25 000	60 000	8 000	2 000	5 000
Kilowatt hours	4 000	18 000	8 200	800	3 300

(a) Calculate, to two decimal places, a suitable overhead absorption rate for each of the three production departments.

Additional information:

The company has been approached with an order from a customer. It is expected that the order will require 11 kilograms of material at $7.80 per kilogram and 6 hours of direct labour at $9.40 per hour. The order should take 5 hours in the cutting department, 4 hours in the machining department and 3 hours in the finishing department. The company's policy is to achieve a profit margin of 50 per cent on all orders.

(b) Calculate the selling price for the order.

Additional information:

In January 2017 it was discovered that in December 2016, 8 800 machine hours were used in the cutting department and 7 100 labour hours were used in the finishing department. All other details were as budgeted.

(c) Calculate the total under-absorption or over-absorption of overheads for December 2016 arising from the actual hours used in the cutting and finishing departments.

Key topics

➤ marginal costing

➤ decision making

➤ comparing traditional methods of costing

➤ business planning and budgeting

➤ greatest challenges

➤ important techniques.

✓ What you need to know

This unit is concerned with: marginal costing including break even analysis; reconciling profit from both marginal costing and absorption costing; the effect of limiting factors on production; cost-volume-profit analysis; unit, job and batch costing statements; accounting and business planning; advantages and disadvantages of budgetary control.

10.1 Marginal costing

Marginal costing is a technique which keeps a clear distinction between variable costs and fixed costs to produce a marginal cost. It can be used to calculate the cost of producing one extra unit.

The technique is applied to finding:

Contribution

Contribution is the difference between selling price and variable costs; it provides the funds available to cover fixed costs.

Break-even point

The relationship between costs and revenue can be illustrated using a break-even chart. Break-even charts are of limited benefit because they are based on assumptions which may unrealistic for the following reasons:

➤ assumes no inventories will be held

➤ the possibility of stepped costs is ignored

➤ assumes variable costs will remain perfectly linear

➤ assumes that the sales line will also be totally linear (so changes in prices for seasonal sales are ignored)

➤ break-even charts are prepared on the basis of a single product being produced, which is not a very common situation.

> **Key term**
>
> **Marginal cost**: the total of all the variable costs required to make one extra product, i.e. the total of variable costs per unit.

> **★ Exam tip**
>
> An understanding of contribution is fundamental to understanding and applying marginal costing techniques.

> **Key term**
>
> **Break-even point**: the point at which total revenue equals total costs, i.e. where there is neither a profit nor a loss being made.

Margin of safety

The margin of safety indicates the extent to which sales could fall before the business will begin to make a loss. The margin of safety can also be expressed as percentage: margin of safety (units)/total sales (units) × 100.

Contribution/sales ratio

measures the relationship between contribution and sales and is expressed as a percentage. The ratio is useful for calculating profit at different levels of activity and break-even point.

Profit/volume chart

shows clearly the profit made at different levels of output and could be considered more useful than a break-even chart.

10.2 Decision-making

➤ Whether to **make or a buy** a product: where, on purely financial grounds, continuing to produce would depend on there being a positive contribution.

➤ Whether to take on a **special order**: again, on purely financial grounds, a special order would be taken on if it made a positive contribution.

➤ **Limited resources**: where a business is faced with a shortage of materials or labour, marginal costing is used to establish the optimum profit that can be made by using the contribution per limiting factor.

Non-financial factors in decision-making

These may need to be considered. For example:

➤ If a business ceases to make a product what will be the impact on the rest of the workforce when there are redundancies?

➤ In a make or buy decision, what will be the quality of the products bought in? Will the supplier be reliable?

➤ When accepting a special order, will production on other products be reduced and if so what will be the effect on regular customers? Will the special order be a one-off or will it lead to further orders? Will staff need to be retrained in order to produce the special order? Will the new customer prove to be reliable?

➤ Maybe a special order should be accepted even though it will have a negative contribution because it will mean that key members of the workforce can be kept on, or will have some charitable effect which will enhance the business's reputation.

10.3 Comparing traditional methods of costing

Marginal costing is particularly effective in a range of situations where critical decisions are to be made; however, the assumption that any particular cost is completely variable or completely fixed is unrealistic in many situations. For example, suppliers may provide greater trade discounts once orders exceed a particular level as production increases, or conversely, a manufacturer may lose trade discounts if production falls and orders drop below a certain level. Absorption costing takes account of all costs and so is suitable for valuing inventory, and does enable any change in costs to be assessed. However, overhead absorption rates are based on machine or labour hours and these may not always be relevant, particularly where modern technology has transformed manufacturing processes.

Revenue statements using marginal costing

The format used below identifies contribution:

	$
Revenue	xx
Less: variable costs	(xx)
Contribution	xx
Less: fixed costs	(xx)
Profit	xx

Cost-volume profit analysis

This is a further development of marginal costing which looks at the effect of differing levels of activity on profit. It is useful for decision-making, planning, cost control, price setting. It suffers, however, from the assumption that fixed costs will remain constant for a particular range of production, and that all costs are either variable or fixed – both of these assumptions can be unrealistic. For example, some costs are semi-variable, and fixed costs may change at a certain production level as when additional storage facilities are required when more goods are produced. It also has its limitations because the assumption is made that selling price will remain constant, ignoring the need for price cutting for seasonal sales.

Unit costing

This is the process of preparing a statement showing each element in the cost of one unit. The statement will include each variable cost per unit plus a fixed cost per unit.

Job costing

This is the process of showing a breakdown of the costs, profit and selling price for a customer's order.

Batch costing

This is the process of showing the costing of a job requiring a quantity of the same product.

10.4 Business planning and budgeting

Planning is one of the most important aspects of business management. It is necessary to ensure that business activities have a clear focus and are based on ideas which will enable the business to survive. Planning is one aspect of the management process, but it is also vital that management has effective procedures for controlling operations requiring the monitoring of actual results against the plan, with action taken when actual results differ from the plan. The planning process is usually broken down into the following stages:

➤ Setting objectives.

➤ Considering different strategies which could be used to achieve the objectives.

➤ Evaluating the strategies, discarding any which are impractical.

➤ Preparing a long-term plan which is expressed in financial terms including a master budget.

➤ Implementing the long-term plan, which involves breaking down the master budget into shorter-term plans/budgets.

Budgets take account of the various constraints for example the availability of: finance, labour, capacity and time.

Benefits of budgeting

Benefits include:

➤ **Planning**: budgets can only be prepared when owners and senior managers have carefully thought out a business strategy, i.e. the future goals, objectives and direction for the business, leading to the allocation of resources.

➤ **Co-ordinating**: encourages communication between managers at each level in an organisation and within teams in each department.

➤ **Communication**: ensuring that managers provide the necessary information for planning and that their ideas do not contradict one another or fail to meet the business's objectives.

➤ **Clarifying responsibilities**: managers will find their responsibilities more sharply defined and normally they will have targets to achieve.

➤ **Control and evaluation**: the performance of managers can be judged by comparing actual results with goals set in budgets. Variances can be identified and judged as controllable (within the control of the manager responsible) or non-controllable (arising from factors beyond the manager's control). The identification of adverse variances provides an opportunity for remedial action to be taken.

➤ **Motivation**: it is often found that individuals become more highly motivated where they have been involved in the budgetary process and where the budgets are challenging but achievable.

Limitations of budgeting

Limitations include:

➤ Reliance on the **accuracy** of estimates: the success of budgetary control is dependent on the accuracy of forecasts.

➤ Budgets set may be **unrealistic**: budgets could be too demanding and unachievable, leading to loss of confidence in the budgeting process and staff demotivation because they are unable to achieve the targets set through no fault of their own. On the other hand, budgets could be undemanding, leading to inefficiencies and under-achievement.

➤ Budgets can be too **restricting** so that managers fail to take up potentially advantageous opportunities because they have not been planned for in a budget.

➤ Budgets can cause **conflict** between departments.

➤ Budget preparation requires **specialist skills** which may not be available especially in a small business.

➤ The process of preparing budgets is **time consuming** and potentially costly.

10.5 What are the greatest challenges?

➤ **Marginal costing:** understanding the concept of contribution.

➤ **Marginal costing:** using break-even analysis and the contribution/sales ratio in problem solving.

➤ **Marginal costing:** decision-making where there are limited resources.

➤ **Budgeting:** understanding the budgeting process and the benefits and limitations of budgetary control.

➤ Writing effective responses to situations where decisions are to be made including consideration of non-financial factors.

10.6 Review some important techniques

Preparing a break-even chart

The steps involved in preparing a break-even chart are:

1. Prepare vertical axis to accommodate the highest figure for sales revenue; prepare the horizontal axis to accommodate the maximum production and sales in units.

2. Label the two axes.

3. Calculate the break-even point and plot this point on the graph.

4. Draw a straight line (it will be horizontal) to represent fixed costs.

5. Draw a straight line to represent sales revenue (it will begin at the origin, 0,0) and pass through the break-even point.

6. Draw a straight line to represent total costs. It will start where the fixed cost line begins on the vertical axis and pass through the break-even point.

7. Label the lines and the break-even point, and indicate the area of profit and the area of loss.

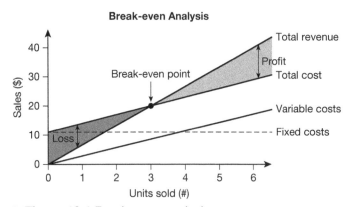

▲ **Figure 10.1** Break-even analysis

> **✗ Common error**
>
> Make sure you label each axis. Don't be caught out by this basic error.

Using break-even analysis and contribution/sales ratio in calculations

> **▼ Example 10.1** Calculating a target profit
>
> The directors of W Ltd would like to achieve a target profit level of $240 000 in 2017. Currently each unit is sold for $80, variable costs are $45 per unit and fixed costs total $530 000. Calculate the number of units which must be sold in 2017 in order to achieve the target profit.
>
> The formula for the break-even point can be adapted in order to make this calculation as follows:

Formula for break-even point	Adaptation to find target profit
$\dfrac{\text{Fixed costs}}{\text{Contribution per unit}}$ = BEP in units	$\dfrac{\text{Fixed costs + Target profit}}{\text{Contribution per unit}}$ = units required to find target profit

So for W Ltd:

$$\frac{\$530\,000 + \$240\,000}{\text{Contribution } \$35 \text{ (i.e. } \$80 - 45)} = 22\,000 \text{ units}$$

Key term

Contribution per unit: selling price per unit less variable costs per unit.

💡 **Remember**

When calculating a break-even point, always round up to the nearest whole unit. It is a common error not to do this. For example, a BEP of 828.21 units should be expressed as 829 units (not 828 units).

▼ **Example 10.2** Using contribution/sales ratio to calculate profit

Using the data in Example 10.1 concerning W Ltd. This company has a contribution/sales ratio of:

$$\frac{\text{Contribution}}{\text{Sales}} \times 100 \text{ i.e. } \frac{35}{80} \times 100 = 43.75\% \text{ (and fixed costs of } \$530\,000)$$

So, if the company achieved sales of $1\,600\,000$, the ratio provides a quick means of calculating profit:

	$
Contribution (43.75% × Sales $1\,600\,000$)	700 000
Less fixed costs	(530 000)
Profit	170 000

Preparing a profit/volume chart

Profit is measured on the vertical axis and output is measured on the horizontal axis.

▼ **Example 10.3** Preparing a profit/volume chart

A company can produce a maximum of 10 000 units a month. Each unit sells for $5 and variable costs are $3 per unit. Fixed costs are $12 000 per month.

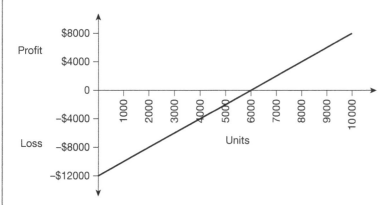

▲ **Figure 10.2** Profit/volume chart

The chart shows that where output is zero the company will incur a loss of $12 000 (i.e. fixed costs), that break-even point will occur when 6000 units are produced and sold; and that the maximum profit per month is $8000 when 10 000 units are produced and sold.

Make or buy decisions

Situations where a manufacturer must decide whether to manufacture a product or buy the product in from elsewhere. On purely financial grounds the decision to make or buy will depend on comparing contributions.

▼ **Example 10.4** Make or buy decision

G Ltd manufacturers product X. Details concerning one unit of product X, based on an annual production of 80 000 units, are as follows:

	$
Materials	6.00
Direct labour	3.50
Other variable costs	1.50
Fixed costs	4.00
Total costs	15.00
Profit	5.00
Selling price	20.00

The directors have found that the same product could be purchased from an overseas supplier for just $13 per unit. The Sales Director has recommended buying in the product because he believes profits will increase from $400 000 per annum to $560 000 per annum.

G Ltd should continue to make the product.

Make the product		Buy the product	
	$		$
Revenue (80 000 x $20)	1 600 000	Revenue (80 000 x $20)	1 600 000
Less Variable costs (80 000 × $11)	(880 000)	Less purchase price (80 000 × $13)	(1 040 000)
Total contribution	720 000	Total contribution	560 000
Less fixed costs (80 000 × $4)	(320 000)	Less fixed costs (80 000 x $4)	(320 000)
Profit	**400 000**	Profit	**240 000**

The comparison demonstrates the importance of comparing contributions and emphasises the fact that fixed costs still have to be covered and they make some time to scale down. There could be other factors to consider if the product was bought in.

➤ Could the factory plant be rented out?

➤ What would be the cost of redundancy payments as existing staff are laid off?

➤ If the factory remains unused will there be extra costs to ensure machinery is kept in working order and the premises remain secure?

➤ Will the supplier be reliable and will the quality of the products be of the usual quality?

 Exam tip

In many questions it will be important to point out and explain other factors for consideration as well as to provide calculations of potential profit.

Accepting a special order

Where there is spare capacity a special order is worth considering, as long as it makes a contribution, as this will help cover fixed costs, or, if break-even point has already been reached, the contribution will increase profits. However, other factors need to be considered too:

➤ Is the new customer reliable – is there a risk that payment will be delayed or not made at all?

➤ Will the special order be instead of other potential orders with a higher contribution?

➤ Will the customer always expect the reduced price should further orders be made?

➤ Will other customers become aware of the special price and expect similar treatment?

Limited resource decisions

When a factor of production (usually materials or labour) will have limited availability for part of a production period, the technique for finding the best outcome for the business in terms of profit is as follows:

1 Calculate the contribution for each type of product.

2 Calculate the contribution per $1 of the scarce resource (i.e. contribution per $1 of materials or contribution per $1 of direct labour).

3 Rank the order for production based on the products with the highest contribution per $1 of the scarce resource.

> **X Common error**
>
> To base the decision on contribution rather than contribution per $1 of the scarce resource.

▼ **Example 10.5** Optimum production plan when materials are in short supply

K Ltd produces three products which each use the same type of material. During March 2017 it has become apparent that materials will have limited availability.

Information about the three products is as follows:

	Product X	Product Y	Product Z
Selling price per unit	30	38	45
Direct materials per unit	5 kg	6 kg	8 kg
Direct labour per unit	2 hrs	2.5 hrs	3 hrs
Budgeted production for March 2017	400	300	200

> **X Common error**
>
> Based on contribution alone the rank order is Z, Y and X – but this ignores the critical element of direct materials per unit in the decision.

Materials cost $2.50 per kg; direct labour rate is $4.80 per hour.

		Product X		Product Y		Product Z	
		$	$	$	$	$	$
Step 1: calculate the contribution per unit	Selling price		30.00		36.00		45.00
	Variable costs						
	Materials	12.50		15.00		20.00	
	Labour	9.60		12.00		14.40	
			(22.10)		(27.00)		34.40
	Contribution		**7.90**		**9.00**		**10.60**

		Product X	Product Y	Product Z
Step 2: Calculate the contribution per $1 of materials	Contribution per unit / Materials per unit	$\dfrac{7.90}{12.50} = \$0.632$	$\dfrac{9.00}{15} = \$0.600$	$\dfrac{10.60}{20} = \$0.530$
Step 3: Place the types of product in rank order		1st	2nd	3rd

Comparing profits using marginal costing and absorption costing

Marginal costing will value inventory at variable cost and absorption costing will value inventory at full cost, as a result profit will be different for each method.

▼ **Example 10.6** Comparing profits using the two traditional costing methods

V Limited produce a single unit. The following details are available:

	$ per unit
Selling price	90
Variable costs	60
Production Dept X overheads	4
Production Dept Y overheads	6

During the year ended 31 December 2016, the company produced 10 000 units and sold 8000 units. Fixed costs for the year were $100 000.

Inventory levels during the year were:

1 January 2016	1 000 units
31 December 2016	2 000 units

The variable cost of marking one unit is $60; the total cost of making one unit is $70. Comparative income statements for the year ended 31 December 2016 are as follows:

	Workings	Marginal Costing		Absorption Costing	
		$	$	$	$
Revenue	8000 x $90		720 000		720 000
Opening inventory	1 000 x $60 marginal cost 1 000 x $70 absorption cost	60 000		70 000	
Production	10 000 x $60 marginal cost 10 000 x $70 absorption cost	600 000		700 000	
Closing inventory	2 000 x $60 marginal cost 2 000 x $70 absorption cost	(120 000)	(540 000)	(140 000)	(630 000)
CONTRIBUTION			180 000		
Fixed costs			(100 000)		
PROFIT FOR YEAR			80 000		90 000

The difference in the reported profit figures arises from the fact that inventories are valued to include fixed costs in the case of absorption costing, but do not include fixed costs in the case of marginal costing.

It is possible to reconcile the difference in reported profits as follows:

	$
Marginal costing profit	80 000
Add fixed cost included in the increase in the closing inventory using absorption costing (increase in inventory is 1000 units x fixed costs per unit $10)	10 000
Absorption costing profit	90 000

Revision checklist

I can:

➤ calculate contribution per unit, margin of safety, break-even point and the contribution to sales ratio ☐

➤ prepare a break-even chart ☐

➤ describe the limitations of break-even analysis ☐

➤ decide whether to make or buy a product ☐

➤ decide whether or not to accept a special order ☐

➤ prepare a production plan to take account of a limited resource ☐

➤ make decisions and provide details of other factors which should be taken into account ☐

➤ calculate profit using marginal costing and absorption costing methods and reconcile the difference. ☐

 Raise your grade

W Ltd have under-used capacity in their factory where an air-conditioning unit is made. Currently 4 800 units are made and sold each year which produces a profit of $100 000 per annum. In normal working 5 400 units could be produced. The summarised results for last year are as follows.

		$
Revenue	$210 per unit	1 008 000
Direct materials	$50 per unit	(240 000)
Direct labour	$65 per unit	(312 000)
Fixed costs		(356 000)
		100 000

The directors require an annual profit of $140 000 and they have been considering the following alternative courses of action.

Option 1: reduce the selling price by 6 per cent which should lead to an increase in sales of 20%. The increase in direct materials required should make it possible to negotiate an additional trade discount of 10%. Overtime working will be required to produce some of the additional units. In overtime conditions, direct labour costs would increase by 15 per cent.

Option 2: replace some outdated machinery and improve production efficiency. This would increase fixed costs by annual finance charges of $4000 and increased depreciation of $15000. However, the new machinery should reduce material wastage resulting in a decrease of 8 per cent in material costs, and require less direct labour resulting in a decrease of 15 per cent in labour costs.

Prepare statements showing the forecast profit for both options and to advise the directors which option should be chosen giving reasons.

Student answer

OPTION 1

	$
Revenue ❷	1137024
DM ❶ ❷	(259200)
DL ❷ ❸	(374400)
Fc	(356000)
❶	147424

OPTION 2

	$
Revenue	1008000
DM ❶ ❷	(220800)
DL ❷	(265200)
Fc ❷	(315000)
❶	147000

The directors should choose Option A as this will provide the greatest profit, though both options achieve the target profit. ❹

How to improve this answer

❶   The answer is not well presented – there is much use of abbreviations and the final result for each option is not labelled.

❷   No workings are provided to support the calculations.

❸   The detail about overtime working has been overlooked resulting in an incorrect answer.

❹   Very limited response; there is not attempt to provide details of other factors which should be considered.

Model answer

OPTION 1

		$
Revenue	5760 **(W1)** × $197.40 **(W2)**	1137024
Direct materials	5760 × $45 **(W3)**	(259200)
Direct labour	5400 × $65	(351000)
	360 **(W4)** × $74.75 **(W5)**	(26910)
Fixed costs		(356000)
	Forecast profit	143914

W1	Increase in volume is 120% × 4800 = 5760
W2	New selling price is 94% × $210 = $197.40
W3	New material price is 90% × $50 = $45
W4	Overtime units: 5760 − 5400 = 360
W5	Overtime labour rate: 115% × 65 = $74.75

OPTION 2

		$
Revenue	unchanged	1 008 000
Direct materials	92% × $240 000	220 800
Direct labour	85% × $312 000	265 200
Fixed costs	**(W1)**	375 000
	Forecast profit	147 000

> W1 Fixed costs:
> $356 000 + finance $4 000
> + $15 000 depreciation

The directors should adopt Option 2. Although both options exceed the profit target, this option will provide the larger figure for profit.

However, the directors should consider other factors before making a final decision:

In the case of both options:

➤ How reliable are the estimates for changes in costs?

In the case of Option 1:

➤ Is the forecast growth in sales reliable?

➤ Are staff prepared to work in overtime?

➤ Will the reduction in price be adopted by competitors, reducing demand in future years?

In the case of Option 2:

➤ Could other increases in fixed costs result from the changes: training on the new machinery; redundancy payments because less direct labour required?

➤ What is the likelihood of the company being able to arrange the finance required for the purchase of new machinery?

➤ Will demand be maintained in future? If not the company could find it difficult to reduce the now more significant fixed costs, leading to a sharp fall in profits.

★ Exam tips

Key features of the model answer

➤ Each element in the answer is carefully labelled and full details supplied.

➤ Workings are provided for each complex calculation.

➤ The answer takes account of all the points given in the question (including the overtime working required for Option 1).

➤ A full answer is given concerning the recommendation including detailed information about other points for consideration. (Note: there are other creditable points which could have been made.)

Exam-style questions

1 Define the term 'contribution'.

2 Describe two limitations of using a system of budgetary control.

3 Q Ltd makes a single product. The following details are available for a production level of 5000 units.

	$ per unit
Selling price	16
Direct materials	4
Direct labour	3
Other variable costs	1
Fixed cost	5

(a) Calculate in units: **(i)** break-even point **(ii)** margin of safety.

The directors believe that by reducing prices by $2 per unit, demand will increase and the company will be able to achieve a target profit of $50 000 per annum. However, it will also be necessary to advertise the product more effectively, so the plan is to spend an additional $500 per month on marketing.

(b) Calculate the level of demand required in order to achieve the target profit.

4 G Ltd, a manufacturing company, has some unused capacity at one of its factories. The company manufactures one product, a 'grix', at this factory. The following details are available for the production of 18 000 units of grix.

	$ per unit
Selling price	82
Direct materials	27
Direct labour	21
Other variable costs	7
Fixed costs	9

A potential customer has offered to buy 1500 units of grix but at a reduced price of $65 per unit.

Advise the directors of G Ltd whether or not they should accept the order from the potential customer.

5 D Ltd manufacture three different models of barbecue: basic, super and deluxe. Planned production for March 2017 is 2000 units of each model. However, it now appears that only

27 000 labour hours will be available during March 2017. The following information is available:

	Basic	Super	Deluxe
Selling price per unit ($)	149.40	195	257.40
Direct materials per unit (kg)	5	6	8
Labour hours per unit	4	6	7

Materials cost $11 per kg; labour costs $10 per hour. Fixed costs per month are $27 000.

In March 2017 the company is committed to producing 500 units of each model for its major customer.

(a) Prepare a statement to show the level of production for each product that would maximise profit in March 2017, taking account of the requirement to supply the major customer with 500 units of each model.

(b) Calculate the profit that will be made based on this plan.

6 K Limited produce a single product. The following details are available:

	$ per unit
Selling price	50
Direct materials	5
Direct labour	7
Other variable costs	3
Machining department overheads	4
Assembly department overheads	8

During the year ended 31 December 2016 the company produced 25 000 units and sold 23 500 units. Inventory levels were:

1 January 2016	500 units
31 December 2016	2 000 units

(a) Prepare comparative income statements to show the profit for the year ended 31 December 2016 using marginal costing and absorption costing.

(b) Prepare a statement to reconcile the differences in the profit from marginal costing and absorption costing.

✓ **What you need to know**

How to prepare a manufacturing account including profit on transfer from factory to finished inventory and accounting for manufacturing profit and the elimination of unrealised profit from unsold inventory.

11.1 The manufacturing account

The purpose of a manufacturing account is to provide a figure for the cost of producing finished goods which can be used in the business's income statement instead of 'purchases' as part of the process of calculating gross profit. A basic manufacturing account has three sections:

▼ **Table 11.1** The manufacturing account

Section	Main content	Details
Prime cost	Total of direct cost (i.e. costs which can be attributed to a single product)	Raw materials consumed
		Direct labour
		Other direct costs
Factory overheads	Total of indirect costs (i.e. all factory costs which cannot be identified with the product)	Factory power
		Factory rent
		Depreciation of factory non-current assets
		Factory manager's salary
		Factory supervisors' salaries
		Factory insurance
		Other indirect costs
Work in progress	Adjustment for the value of work in progress	Opening inventory of work in progress less the closing value of work in progress

Key terms

Raw materials consumed: the cost of raw materials used during a financial period based on the opening inventory of raw materials plus purchases (add carriage inwards, less returns outwards) less the closing inventory of raw materials.

Prime cost: the total of direct costs of producing goods.

The costs within a manufacturing account are added to find the overall total "cost of production", i.e. prime cost + subtotal for factory overheads +/– adjustment for work in progress.

11.2 The income statement of a manufacturing organisation

The first section leading to the gross profit features the revenue of the business less the cost of sales. This section is entirely concerned with finished goods.

11.3 The statement of financial position of manufacturer organisation

This could include three separate inventory figures: raw materials, work in progress and finished goods.

11.4 Manufacturing profit

In some manufacturing organisations finished goods are transferred from the manufacturing account to the income statement at a transfer price. This is a manufacturing profit. The transfer price is usually based on how much the goods would have cost had they been bought in. The result is to split off part of the profit for the year and record this in the manufacturing account. The reason for this procedure is to identify how much profit is attributable to the production element of the organisation's activities and how much profit is attributable to the trading element of the organisation's activities.

11.5 Provision for unrealised profit

Where a manufacturing organisation chooses to identify manufacturing profit, inventories of finished goods will include an element of this factory profit, i.e. they will be valued at transfer price. In order to eliminate this profit element from the value of inventories of finished goods a provision for unrealised profit is required. The provision ensures that inventories are recorded in the statement of financial position at cost price. The income statement records the initial creation of the provision and thereafter annual adjustments to the figure made necessary because inventories of finished goods have increased or decreased. In effect, the provision for unrealised profit is recorded in the same way as a provision for doubtful debts.

11.6 Accounting concepts and manufacturing organisations

As well as the application of concepts common to the preparation of all financial statements, the recording of a provision for unrealised profits results from applying the:

➤ **prudence concept**: to ensure that the value of inventories of finished goods are not overstated resulting in an overstatement of profit

➤ **realisation concept**: to ensure that profit is only recognised when goods are sold (i.e. the production of finished goods does not guarantee that a profit will be made).

11.7 What are the greatest challenges?

➤ Selecting the correct information to be included in each section of a manufacturing account.

➤ Correctly recording manufacturing profit.

➤ Calculating and correctly recording adjustments to the provision for unrealised profit.

11.8 Review some important techniques

Preparing a manufacturing account to include a manufacturing profit.

▼ **Example 11.1** L Manufacturers Ltd. is preparing financial statements for the year ended 31 December 2016. The following information is available:

	$
Carriage inwards on raw materials	4 500
Depreciation of factory machinery	22 800
Direct labour	137 000
Indirect labour	27 900
Inventories at 1 January 2016	
Raw materials	11 400
Work in progress	8 600
Inventories at 31 December 2016	
Raw materials	13 200
Work in progress	9 300
Other factor overheads	29 600
Purchases of direct materials	82 400
Returns outwards	1 700

The company transfers finished goods from the factory to the income statement at cost plus 25 per cent.

Manufacturing account for the year ended 31 December 2016

	$	$	$
Cost of raw materials consumed			
Opening inventory		11 400	
Purchases	82 400		
Carriage inwards	4 500		
	86 900		
Returns outwards	(1 700)		
		85 200	
		96 600	
Closing inventory		(13 200)	
			83 400
Direct labour			137 000
Prime cost			220 400
Factory overheads			
Depreciation of factory machinery		22 800	
Indirect labour		27 900	
Other overheads		29 600	
			80 300
			300 700
Adjustment for work in progress			
Opening inventory		8 600	
Closing inventory		(9 300)	
			(700)

Production cost of manufactured goods	300 000
Manufacturing profit (at 25%)	75 000
Transfer price of finished goods	375 000

Common error

Check that you have added factory overheads to the prime cost. It is a common mistake to deduct the factory overheads rather than add them.

Calculating changes in the provision for unrealised profit

Each year the provision for unrealised profit on finished goods must be adjusted to keep in step with increases or decreases in the closing inventory.

▼ **Example 11.2** L Manufacturers Ltd. (see Example 11.1 on previous page) have provided the following information:

	$
Inventory of finished goods at 1 January 2016	20 000
Inventory of finished goods at 31 December 2016	16 000

It is important to realise that the figures provided will be at transfer price, so they both include unrealised profit based on increasing the cost value of the inventories by 25 per cent.

The provision for unrealised profit at 1 January 2016 must be $4000; and the cost of the finished goods at this date have been $16 000.

Check: cost $16 000 plus increase of 25 per cent in cost (i.e. 25% × $16 000) is $4000 = $20 000.

So the procedure to find the provision for unrealised profit is:

$$\frac{\text{Inventory at transfer price} \times 25}{100 + 25} \quad \text{i.e.} \quad \frac{20\,000 \times 25}{125} = \$4000$$

The provision for unrealised profit required on the closing inventory is:

$$\frac{\text{Inventory at transfer price} \times 25}{100 + 25} \quad \text{i.e.} \quad \frac{16\,000 \times 25}{125} = \$3200$$

So at 31 December 2016 the provision or unrealised profit should be reduced by $800.

In the statement of financial position, the current assets section will include the following details:

Statement of financial position at 31 December 2016 (extract)

Current assets	$	$
Inventories:		
Raw materials		13 200
Work in progress		9 300
Finished goods at transfer price	16 000	
Less provision for unrealised profit	3 200	
		12 800
		35 300

Recording manufacturing profit and adjustments to the provision for unrealised profit in the income statement

▼ **Example 11.3** An income statement for L Manufacturers Ltd. can be prepared using the information in illustrations 1 and 2 and the following details:

	$
Office and distribution expenses	173 700
Revenue	540 000

Income statement for the year ended 31 December 2016

	$	$
Revenue		540 000
Less cost of sales (at transfer price)		
Opening inventory	20 000	
Production of finished goods	375 000	
	395 000	
Closing inventory	(16 000)	
		(379 000)
Gross profit		161 000
Add manufacturing profit	75 000	
less decrease in provision for unrealised profit	(800)	
		74 200
		235 200
Office and distribution expenses		(173 700)
Profit for the year		61 500

Revision checklist

I can:

➤ define and explain key terms such as prime cost, factory overheads, manufacturing profit, provision for unrealised profit ☐

➤ explain why some manufacturing organisations record a manufacturing profit and the potential benefit from doing so ☐

➤ explain how accounting concepts are applied to inventories valued at transfer price ☐

➤ prepare manufacturing accounts including manufacturing profit ☐

➤ calculate changes in the provision for unrealised profit ☐

➤ prepare income statements to include manufacturing profit and adjustments to the provision for unrealised profit. ☐

Raise your grade

H Ltd. are manufacturers. The following information is provided for the year ended 31 December 2016.

	$
Carriage outwards	4 000
Direct labour	140 000
Factory overheads	50 000
Inventories 1 January 2016	
Raw materials	18 000
Work in progress	11 000
Finished goods	28 000
Inventories 31 December 2016	
Raw materials	24 000
Work in progress	15 000
Finished goods	21 000
Office expenses	27 000
Provision for unrealised profit 1 January 2016	8 000
Purchases of raw materials	120 000
Revenue	485 000

Finished goods are transferred from the factory at cost plus 40 per cent.

Prepare a manufacturing account and income statement for the year ended 31 December 2016.

Student answer

Manufacturing account for the year ended 31 December 2016

	$	$
Cost of raw materials consumed		
Opening inventory	18 000	
	120 000	
Carriage outwards ❶	4 000	
	142 000	
Closing inventory	(24 000)	
		118 000
Direct labour		140 000
		258 000
Factory overheads ❷		(50 000)
❽		208 000
Adjustment for work in progress ❸		
Opening inventory	11 000	
Closing inventory	15 000	
		4 000
Cost of production		212 000
Manufacturing profit (40%)		84 800
Transfer price of manufactured goods		296 800

Income statement for the year ended 31 December 2016

	$	$
Revenue		485 000
Cost of sales at transfer price		
Opening inventory	20 000	
Production of finished goods	296 800	
	316 800	
Closing inventory	16 000	
		300 800
Gross profit ④		184 200
Decrease in provision for unrealised profit ⑤ ⑥ ⑦	6 400	
Office expenses	27 000	
		(33 400)
⑧		150 800

How to improve this answer

The common errors made by the candidate are:

① Carriage outwards has been misidentified as a direct cost; carriage outwards is a distribution cost and should have been included in the income statement.

② The factory overheads (indirect costs) have been deducted from prime cost.

③ The adjustment for work in progress should have been deducted.

④ The candidate has forgotten to add manufacturing profit to gross profit.

⑤ The provision has been miscalculated as 40 per cent of closing inventory rather than 40/140 of closing inventory.

⑥ Only the decrease in the provision should be recorded.

⑦ If the provision decreases, it should be added back to profit not deducted.

⑧ Some important labels have been omitted: prime cost, profit for the year.

Model answer

Manufacturing account for the year ended 31 December 2016

	$	$
Cost of raw materials consumed		
Opening inventory	18 000	
Purchases	120 000	
	138 000	
Closing inventory	(24 000)	
		114 000
Direct labour		140 000
Prime cost		254 000
Factory overheads		50 000
		304 000
Adjustment for work in progress		
Opening inventory	11 000	
Closing inventory	(15 000)	
		(4 000)

Cost of production		300 000
Manufacturing profit (40%)		120 000
Transfer price of manufactured goods		420 000

Income statement for the year ended 31 December 2016

	$	$
Revenue		485 000
Cost of sales at transfer price		
Opening inventory	28 000	
Production of finished goods	420 000	
	448 000	
Closing inventory	(21 000)	
		(427 000)
Gross profit		58 000
Add manufacturing profit	120 000	
Add decrease in provision for unrealised profit **(W1)**	2 000	
		122 000
		180 000
Carriage outwards	4 000	
Office expenses	27 000	
		(31 000)
Profit for the year		149 000

	$
W1 Decrease in provision for unrealised profit	
Provision at 1 January 2016	8 000
Provision at 31 December 2016*	(6 000)
	2 000

*Provision is $21 000 × 40/140 = $6 000

1 State one reason why some manufacturing businesses record a manufacturing profit in their financial statements.

2 Explain why it is necessary to make a provision or unrealised profit when finished goods are transferred from the factory at transfer price.

3 Calculate the provision for unrealised profit in each of the following situations.

Inventory at transfer price	Percentage added to cost to achieve transfer price
$	%
24 000	20
18 000	25
36 000	33 1/3
48 000	60

4 N Ltd., a manufacturing company, has provided the following information for the year ended 30 September 2016.

	$
Carriage outwards	3 000
Depreciation factory machinery	14 900
Depreciation office equipment	4 700
Factory power	3 700
Factory labour	190 000
Insurance	4 800
Inventories 1 October 2015	
raw materials	16 280
work in progress	13 200
finished goods	12 600

	$
Inventories 30 September 2016	
raw materials	15 770
work in progress	11 200
finished goods	15 000
Office salaries	22 480
Provision for unrealised profit	2 100
Purchases of raw materials	97 490
Rent	13 800
Revenue	585 000

Additional information:

- Finished goods are transferred from the factory at cost plus 20 per cent.
- Rent $600 was due but unpaid at 30 September 2016.
- Factory labour is 90 per cent direct; the remainder is indirect.
- Insurance and rent should be allocated: 75 per cent factory; 25 per cent office.

Prepare:

(a) A manufacturing account for the year ended 30 September 2016.

(b) An income statement for the year ended 30 September 2016.

(c) An extract from the statement of financial position at 30 September 2016 showing how inventories should be recorded.

Key topics

➤ not-for-profit accounts

➤ financial statements

➤ greatest challenges

➤ important techniques.

✓ What you need to know

How to prepare a trading account, income and expenditure account and statement of financial position. How to make adjustments to expense and income items including members' subscriptions; and how to record donations and life membership funds.

12.1 Not for profit accounts

Not-for-profit organisations cover a wide range of entities including clubs, societies and charities. These organisations differ from businesses in that they have quite different objectives. Their main purpose is to provide a service for their members, rather than to trade and make a profit for owners. (Please note that for convenience references here are to clubs, but all the points made could apply to the whole range of not-for-profit organisations.)

Although profit is not the main objective, accounting information has a very important role to play in ensuring, for example, that a club's members are aware of the club's financial performance, so that they can better understand and give their opinion on the key decisions made by the club's treasurer and committee to ensure that the club is financially viable.

12.2 Financial statements

There are various financial statements that a not-for-profit organisation can use:

Receipts and payments account

A receipts and payments account shows the change in the club's cash resources (both cash and bank balances combined) during the year under review. The account will record:

➤ the opening balance(s) of cash in hand and/or cash at bank

➤ all receipts (capital and revenue)

➤ all payments (capital and revenue)

➤ the closing balance(s) of cash in hand and/or cash at bank.

Accounts to show the profit or loss on activities

These accounts are designed to identify additional income for the club. They can be:

➤ relatively simple – possibly for a one-off event – in which case the details can be presented within the income and expenditure account and show a profit or loss on the activity

Key term

Receipts and payments account: a summary of all the detailed information in a club's cash book.

A Level

➤ more complex – possibly involving trading activities and possibly taking place throughout the year – in which case a separate account will be required to include all details relating specifically to that activity and concluding with the profit or loss.

Income and expenditure account

An income and expenditure account is designed to show whether the club's income is sufficient to cover the club's costs. It is broadly similar to the income statement prepared for a business organisation.

➤ Income could include membership subscriptions, profit made on activities, profit on the disposal of non-current assets, donations.

➤ Expenditure could include running costs including depreciation of non-current assets, losses on activities, loss made on the disposal of non-current assets, secretary's and/or treasurer's honorarium (i.e. a fee paid to reimburse the individual for expenses paid on behalf of the non-profit making organisation).

The account will either show a surplus for the year or a deficit for the year. The terms profit or loss should not be used as the statement is for a not-for-profit organisation.

Statement of financial position

This is prepared on similar lines to those for business organisations to show assets and liabilities, but will show a figure for **'accumulated fund'** instead of capital. The accumulated fund is the net value of the club which will either increase (when there is a surplus) or decrease (when there is a deficit).

12.3 What are the greatest challenges?

➤ Absorbing a considerable amount of detail. Questions on this topic are likely to contain a considerable amount of information which tests the skills of the candidate in being able to assimilate information quickly and select the right information at the right moment.

➤ There are various technical issues which require a good understanding of accounting principles and procedures for example:

  ➤ Distinguishing between revenue items (required for the income and expenditure account and supporting accounts) and capital items (such as loans, non-current assets, required for the statement of financial position).

  ➤ Making accurate calculations of income from subscriptions or expenses where there are multiple adjustments to be made.

12.4 Review some important techniques

Making adjustments

The income and expenditure account is prepared on an accruals basis, so it is necessary to make adjustments as follows:

➤ Receipts from members' subscriptions for:

  ➤ opening balances of subscriptions due and subscriptions received in advance

  ➤ closing balances of subscriptions due and subscriptions received in advance.

➤ Payments for expenses for:

➤ opening balances of expenses due and expenses prepaid

➤ closing balances of expenses due and expenses prepaid.

★ **Exam tip**

Many students find it helpful to apply the following rule when making adjustments: ADD adjustments detailing information relating to the current year; DEDUCT adjustments detailing information relating to the previous year or the next year.

▼ **Example 12.1** Making multiple adjustments

The treasurer of the Bunker Golf Club has provided the following details:

Extract from Receipts and Payments Account for the
year ended 31 December 2016

$

Receipts

Members' subscriptions   14 500

Payments

Rent of clubhouse   7 300

Additional information

	At 1 January 2016	At 31 December 2016
	$	$
Clubhouse rent due but unpaid	600	
Clubhouse rent paid in advance		300
Subscriptions due	700	200
Subscriptions received in advance	400	500

Calculation of expenditure on clubhouse rent

	$	Guidance notes
Payment	7 300	
Opening balance: rent due	(600)	Deduct the opening balance of rent due as this item is rent for the previous year, i.e. this is rent due for the previous year paid in the current year.
Closing balance: rent paid in advance	(300)	Deduct the closing balance of rent paid in advance as this item is rent for the next year, i.e. this is rent paid in advance for the next year.
Amount to be shown in income and expenditure account	**6 400**	

Calculation of income from subscriptions

	$	Guidance notes
Receipts	14 500	
Opening balance: subscriptions due	(700)	Deduct the opening balance of subscriptions due as this item is subscriptions relating to the previous year. These are subscriptions due for the previous year, but received in the current year.
Opening balance: subscriptions received in advance	200	Add the opening balance of subscriptions received in advance as this item is for subscriptions relating to the current year. These are subscriptions received last year, but which relate to the current year.
Closing balance: subscriptions due	400	Add the closing balance of subscriptions due as this item is for subscriptions relating to the current year. These are subscriptions which will be received next year but which relate to the current year.
Closing balance: subscriptions received in advance	(500)	Deduct the closing balance of subscriptions received in advance as this item is for subscriptions relating to next year. These are subscriptions received in the current year but which relate to the next year.
Amount to be shown in income and expenditure account	**13 900**	

Sometimes it is a requirement to prepare a subscriptions account, which tests understanding of the double-entry model for an income item.

Here is the correct layout for a subscriptions account:

Subscriptions Account

Dr				Cr
	$			$
Opening balance (due)	Asset	Opening balance (received in advance)		Liability
Income and expenditure account	xxx	Receipts		xxx
Closing balance c/d (received in advance)	xxx	Closing balance c/d (due)		xxx
	xxx			xxx
Balance b/d (due)	Asset	Balance b/d (received in advance)		Liability

Calculating the opening accumulated fund

To calculate the accumulated fund, total the value of the club's assets and deduct the total value of the club's liabilities at the beginning of the year.

Key term

Accumulated fund: a fund that represents the value of the club – it is the equivalent of a business's capital.

Donations

Donations received by a club are usually treated as annual income on the grounds that they are likely to be a relatively minor source of income and occur in most, if not every, financial year. However, where the amount is exceptionally large and might be considered a one-off receipt, the club's committee might decide it is more appropriate to treat the donation as a capital item because its inclusion in the income and expenditure could easily distort the income for the year, giving members a rather false and optimistic view of the club's financial viability. Where the donation is to be treated as a capital item, the amount should be shown as a separate item often added to the accumulated fund in the statement of financial position.

★ Exam tip

Remember to include any opening cash and/ or bank balances.

Life membership schemes

A life membership subscription is a one-off, and often substantial, payment made by a member who will then be entitled to use the club's facilities indefinitely. Income from life membership subscriptions should, of course, be spread out over the years the members uses the facilities. However, since this is impossible to determine, the club committee usually devises a policy to spread subscription over a number of years (for example, ten years). Life membership subscriptions are recorded in a life membership fund account from which an annual transfer is made to the club's income and expenditure account.

★ Exam tip

The balance of the account is shown as a separate item in the club's statement of financial position under the subheading non-current liabilities.

Revision checklist

I can:

➤ define and explain the key terms ☐

➤ make complex adjustments applying the accruals concept for expense items and for subscription income ☐

➤ prepare accounts to show the profit or loss on activities including accounts for trading activities and accounts for other activities designed to provide additional income for the organisation ☐

➤ apply the correct accounting treatment for donations and life membership schemes ☐

➤ prepare an income and expenditure account which includes all relevant information, but which excludes items of capital expenditure and capital revenue ☐

➤ calculate a club's accumulated fund at the beginning of a financial year ☐

➤ prepare a statement of financial position showing all relevant information including life membership funds ☐

➤ review the financial performance of a not for profit organisation, providing an analysis of possible courses of action and discuss the benefits and drawbacks of these courses of action. ☐

Raise your grade

The following information relates to the Haven Sports Club.

	At 1 January 2016	At 31 December 2016
	$	$
Accumulated fund	?	?
Administration expenses due	360	510
Insurance prepaid	70	40
Inventory of café supplies	570	760
Life membership fund	3500	?
Loans from members (long-term)	5000	6300
Sports equipment	24600	29200
Subscriptions due	580	660
Subscriptions received in advance	420	290
Trade payables for café supplies	690	480

Receipts and Payments Account

Dr	$		Cr	$
Opening balance of cash at bank	4900	Additional sports equipment		7400
Members' subscriptions	8920	Trade payables for café supplies		8720
Café sales	11490	Administration expenses		2250
Social evening receipts	900	Social evening expenses		510
Donation	1500	Insurance		590
Life membership fund	800	Staff wages		3660
Loans from members (long-term)	1300	Rent of clubhouse and grounds		7200
Closing balance of cash at bank	520			
	30330			30330

Additional information:

The club committee have decided the donation should be regarded as a capital receipt.

Staff wages includes $1100 for café staff.

The club's policy is to consider 15 per cent of the life membership fund as annual income.

Prepare the following:

(a) An account to show the café profit and loss.

(b) An income and expenditure account for the year ended 31 December 2016.

(c) A statement of financial position at 31 December 2016.

[Note: only model answer given for this part]

Student answer

Café account ❶

	$	$
Revenue		11 490
Opening inventory of café supplies	570	
Purchases (8720 + 510 – 760) ❷	8 530	
Less closing inventory of café supplies	(820)	
		(8 380)
❸		3 110
Less wages		(1 100)
❸		2 010

Income and Expenditure Account for the year ended
31 December 2016

	$	$
INCOME		
Subscriptions ❹	8 870	
Social evenings ❺	900	
Loan from members ❻	1 300	
Life membership fund (15% x 3500) ❼	525	
❽		11 595
EXPENDITURE		
Sports equipment ❾	7 400	
Administrative expenses ❿	2 250	
Social evening expenses ❺	510	
Insurance (590 + 70 – 4)	620	
Staff wages	2 560	11 595
Rent of clubhouse and grounds	7 200	
		(20 540)
Loss for year ❸		8 945

How to improve this answer

❶ The title 'café account' is inadequate – it should include reference to the time period.

❷ Workings have been included for the purchases figure, but they are incorrect.

❸ Labelling of key figures has been overlooked in the café account, and the wrong label used in the income statement, i.e. loss for year should be deficit for year.

❹ No workings have been provided to support the subscriptions figure which is incorrect, so all marks for this entry would be lost.

❺ Club members should be given a figure for the profit or loss on the social evenings.

❻ Loans from members are capital receipts not income.

❼ Income from the life membership fund has been miscalculated.

❽ The profit made by the café has been overlooked.

⑨ Equipment has been included which is capital expenditure and depreciation has been overlooked.

⑩ Adjustments to administrative expenses have been overlooked.

Model answer

Haven Sports Club
Café Trading Account
for the year ended 31 December 2016

	$	$
Revenue		11490
Opening inventory	570	
Purchases (**W1**)	8510	
Closing inventory	(760)	
Cost of sales		(8320)
Gross profit		3170
Less: café staff wages		(1100)
Profit for year		2070

W1 Café purchases

	$
Payments to trade payables	8720
Less opening amount due	(690)
Add closing amount due	480
	8510

Income and Expenditure Account
for the year ended 31 December 2016

	$	$	$
Income			
Café profit		2070	
Members' subscriptions (**W2**)		9130	
Profit on social evenings			
receipts	900		
less expenses	(510)		
		390	
Life membership fund (**W3**)		645	
			12235
Expenditure			
Administration expenses			
(2250 − 360 + 510)		2400	
Insurance			
(590 + 70 − 40)		620	
Staff wages (3660 - 1100)		2560	
Rent of clubhouse and grounds		7200	
Depreciation of sports equipment (**W4**)		2800	
			(15580)
Deficit for year			3345

W2 Income from subscriptions

	$
Receipts	8 920
Opening balance due	(580)
Opening balance received in advance	420
Closing balance due	660
Closing balance received in advance	(290)
	9 130

Alternative version set out as an account:

Subscriptions Account

Dr				Cr
	$			$
Opening bal (due)	580	Opening bal (received in advance)		420
Income and expenditure account	9 130	Receipts		8 920
Closing balance c/d (received in advance)	290	Closing balance c/d (due)		660
	10 000			10 000
Bal b/d (due)	660	Bal b/d (received in advance)		290

W3 Life membership fund:

	$
Opening balance	3 500
Receipts during year	800
	4 300
Income for year (15% x 4300)	(645)
Closing balance	3 655

W4 Depreciation of sports equipment

	$
Opening balance	24 600
Additional equipment	7 400
	32 000
Less closing valuation	(29 200)
Depreciation	2 800

Statement of financial position at 31 December 2016

	$	$
Non-current assets		
Sports equipment		29 200
Current assets		
Café inventory	760	

Other receivables		
insurance prepaid	40	
subscriptions due	660	
		1460
Total assets		30660
Accumulated fund		
Opening balance (**W5**)	20750	
Less deficit for year	(3345)	
Donation	1500	
		18905
Non-current liabilities		
Life membership fund	3655	
Loans from members	6300	
		9955
Current liabilities		
Trade payables	480	
Other payables		
administration expenses	510	
subscriptions received in advance	290	
Bank overdraft	520	
		1800
Accumulated fund and liabilities		30660

W5 Opening balance of accumulated fund:

	$	$
Assets		
Insurance prepaid	70	
Inventory of café supplies	570	
	24	
Sports equipment	600	
Subscriptions due	580	
Balance at bank	4900	
		30720
Liabilities		
Administration expenses due	360	
Life membership fund	3500	
Loans from members	5000	
Subscriptions received in advance	420	
Trade payables	690	
		(9970)
		20750

Key features of the model answer

➤ The answer begins with the name of the club and a full title for the café account.

➤ Correct labels are shown throughout for the key subtotals and totals.

➤ The workings for purchases, subscriptions income, income from the life membership fund, equipment depreciation and administrative expenses have been included and are correct.

➤ The café profit has been included in the income and expenditure account.

➤ The profit on the social evenings is clearly identified.

➤ The income and expenditure account does not include any capital receipts (members' loans) or capital expenditure (equipment).

➤ In the statement of financial position subscriptions due has been correctly recorded as a current asset and subscriptions received in advance has been correctly recorded as a current liability; the life membership fund has been shown as a non-current liability, and detailed workings have been given for the opening balance of the accumulated fund.

? Exam-style questions

1 State what is meant by the term 'accumulated fund'.

2 Explain the accounting treatment of a life membership fund.

3 Advise a club treasurer on how a donation should be recorded in the club's financial statements.

4 A club provides a café for the use of its members. The club has been running at a loss for the last few years. The club committee has been considering whether to close the café or, alternatively, make some improvements to the café facilities to make it more attractive for members. Discuss the factors the committee should consider when deciding whether to close the café or update its facilities.

5 The treasurer of the Golden Rally Squash Club has provided the following information about the club's financial year ended 30 September 2016.

	$	$
Balance at bank 1 October 2015		2 920
Receipts		
Members' subscriptions	23 800	
Shop receipts	5 490	
Annual dinner and social evening	2 410	
Donations	3 600	
Competition entry fees	1 230	
Life membership fund	990	
Proceeds from sale of equipment	270	
		37 790
		40 710
Payments		
Suppliers of goods for shop	3 100	
General running costs	6 430	
Rent of club facilities	5 000	
Annual dinner and social evenings costs	1 570	
Competition travel expenses	1 100	
Competition prizes	480	
Purchase of new equipment	6 200	
Transfer to savings accounts	6 800	
Shop staff wages	610	
Secretary's honorarium	300	
		(31 590)
Balance at bank 30 September 2016		9 120

	at 1 October 2015 $	at 30 September 2016 $
Competition travel expenses due	-	340
Equipment	14 800	?
General running costs		
due	-	160
prepaid	220	-
Life membership fund	8 610	?
Members' subscriptions		
due	780	420
received in advance	270	540
Savings account	11 400	?
Shop inventory	1 550	?
Trade payables for shop supplies	420	730

Additional information:

1 The transfer to the savings account was made on 1 April 2016. Interest at 10 per cent per annum is paid annually and added directly to the savings account. Interest for the year ended 30 September 2016 has not yet been received.

2 The treasurer had not had an opportunity to value the shop inventory on 30 September 2016. However, all goods are sold with a gross profit margin of 50 per cent.

3 The club's policy is to consider 12.5 per cent of the life membership fund as annual income.

4 The club committee have decided that 75 per cent of the donations received should be capitalised.

5 The equipment sold had a NBV of $600 on 1 October 2015. Equipment is depreciated by 25 per cent each year, based on the net book value of equipment in use at the year end.

Prepare:

(a) An account to show the profit or loss for the year ended 30 September 2016 for the shop.

(b) An income and expenditure account for the year ended 30 September 2016.

(c) A statement of financial position at 30 September 2016.

Additional information:

The club committee's main aim is to encourage as many people as possible to take up squash as a sport. The committee has been considering moving the club to alternative accommodation. If the move took place, the committee plan to use funds totalling $15 000 from the savings account to purchase additional equipment. The rent of the new accommodation would be 15 per cent higher per annum than that currently paid. A survey has shown that if the club moved to the alternative accommodation and improved the equipment, membership would increase by 20 per cent, assuming there was an increase of only 5 per cent in the annual subscription fee. However, there would not be space available to continue to operate a shop.

(d) Advise the club committee on whether the proposed plans should take place. Give reasons for your advice.

Key topics

➤ statutory requirements

➤ international accounting standards

➤ role of auditors

➤ role of directors

➤ greatest challenges

➤ review of important techniques.

✓ What you need to know

The study of limited companies includes financial statements and the regulatory framework in which they operate; preparing income statements, statements of cash flows and statements of financial position. You will need to be able to explain and apply the main provisions of certain international accounting standards. You will also need to be able to explain the role of the auditor, explain and discuss the role of directors and their responsibilities to shareholders, and discuss the importance of a true and fair view in respect of financial statements.

13.1 Statutory requirements

Limited companies are required to publish their financial statements annually to enable shareholders to assess the performance of the company in which they have invested. The financial statements are sent to each shareholder before the annual general meeting and are also filed with the registrar of companies so that anyone can access their contents. The published financial statements must include:

➤ an income statement

➤ a statement of changes in equity

➤ a statement of financial position

➤ a statement of cash flows

➤ notes to the accounts

➤ a directors' report

➤ an auditors' report.

The notes to the accounts provide more detailed information about content of the income statement and statement of financial position. This is necessary because these two statements are usually published in a somewhat summarised form.

Key term

Statement of cash flows: a statement which highlights how cash funds have been generated during a financial period and how these funds have been utilised.

13.2 International accounting standards

The international accounting standards are designed to protect users of financial statements and to prevent stakeholders being misled. They implement five fundamental accounting concepts:

1 true and fair view

2 duality

3 consistency

4 business entity

5 money measurement.

and are designed to ensure:

➤ comparability

➤ relevance

➤ reliability

➤ understandability.

Key features of international accounting standards

▼ **Table 13.1** International accounting standards

Key term

International accounting standards: agreed rules and guidance to be followed in many countries worldwide which regulate how financial statements are presented to the benefit of stakeholders and users of the statements.

IAS1	**Presentation of financial statements**
	➤ Lists the financial statements that must be prepared by the directors (the same list as in 13.1 on page 137, but without the directors' report and the auditors' report).
	➤ Lists specific concepts that must be applied: going concern, accruals, consistency and materiality.
	➤ Requires the inclusion of comparative figures from previous years.
	➤ Requires notes to the accounts to include details of the company's accounting policies, for example depreciation methods used.
IAS2	**Inventories**
	➤ Requires that inventories should be valued at the lower of cost and net realisable value.
	➤ Defines cost as the purchase price plus any additional cost incurred in order to put the goods in a saleable condition (e.g. carriage).
	➤ Accepts valuations based on FIFO, AVCO and standard cost, but not LIFO.
IAS7	**Statement of cash flows**
	➤ Requires the statement to be divided into three sections: operating activities, investing activities and financing activities.
	➤ Requires a total to be shown for each of these three sections.
IAS8	**Accounting policies, changes in estimates and errors**
	➤ All accounting policies detailed in any of the international standards have to be implemented.
	➤ In other situations, directors must choose which policies are to be followed having regard to the overriding requirement to provide a true and fair view, and to provide reliable, relevant and understandable information.
	➤ Policies should be applied consistently.
	➤ Errors should be corrected when they are discovered, and adjustments made to comparable figures if appropriate.

IAS10	**Events after the reporting period**
	➤ Adjusting events are those events (favourable or unfavourable) reported after the reporting period which provide evidence of conditions that existed during the reporting period. Directors may not authorise the financial statements until they are changed in the light of this evidence.
	➤ Non-adjusting events are those events which have arisen since the end of the reporting period and they are not shown within the financial statements, but if material should be disclosed as notes to the accounts.
	➤ Proposed dividends are regarded as a non-adjusting event.
IAS16	**Property, plant and equipment**
	➤ Covers the treatment of tangible non-current assets described as property, plant and equipment.
	➤ Requires that non-current assets are recorded at cost.
	➤ Cost should take account of the purchase price plus any other costs incurred in order to prepare the asset for use.
	➤ In the financial statements non-current assets should either be recorded at cost less depreciation to provide a net figure called the carrying amount, or at revaluation.
	➤ When a non-current asset is revalued it should be based on the market value of the assets should it be sold (i.e. its fair value). Where property is revalued a professional revaluation is required.
	➤ Maintenance and repair costs are to be recorded as revenue expenditure.
	➤ Tangible assets (except land) are to be depreciated using either the straight-line or reducing balance method, taking account of expected usage, physical wear and tear, obsolescence and legal limits on the use of the asset.
	➤ Derecognition is the term used to cover the situation where an asset is sold or no longer providing any economic benefit.
	➤ A schedule of non-current assets is to be included in the published accounts.
IAS36	**Impairment of assets**
	➤ Defines recoverable amount as the maximum value that can be placed on a non-current asset. It is the higher of an asset's fair value or value in use.
	➤ Fair value is the amount for which an asset could be sold less any selling costs.
	➤ Value in use is the total of estimated future cash flows.
	➤ An impairment loss occurs when the recoverable amount is less than the carrying amount (net book value) and the loss must be shown in the income statement. Usually an impairment loss occurs when a non-current asset has become damaged or is no longer in use.
	➤ All non-current assets must be assessed annually to establish whether there are impairment losses.

IAS37	**Provisions, contingent liabilities and contingent assets**
	➤ Defines liabilities as a present obligation arising from past events which lead to a payment being made.
	➤ Defines provisions as a liability of uncertain timing or amount.
	➤ Defines a contingent liability as: either a possible (less than 50 per cent chance) obligation arising from past events which will be determined by some future event which may or may not occur; or a present obligation that is not probable or cannot be quantified. Where the obligation is possible (rather than a remote possibility) the contingent liability should be disclosed in notes to the financial statements.
	➤ Defines a contingent asset as a possible asset arising from past events that depends on some future event which is not within the entity's control. Where the contingent asset is probable it is shown as a note to the accounts; where it is only possible or remote no disclosure is required.
IAS37	**Intangible assets**
	➤ Defines an intangible asset as an identifiable non-monetary asset without physical substance.
	➤ An intangible asset must be within the organisation's control and be of future economic benefit to the organisation.
	➤ Intangible assets can be purchased or may be internally generated.
	➤ Intangible assets must have an identifiable value if they are to be capitalised.
	➤ Certain internally generated intangible assets must be charged to the income statement including internally generated goodwill, relocation costs, training costs.
	➤ Intangible assets must be valued at cost less accumulated amortisation (i.e. depreciation) or at a revaluation.
	➤ Research expenditure is treated as a revenue item as it would not be possible to know if any economic benefit will arise.
	➤ Development expenditure can be treated as an intangible asset where the project can be completed, the intention is to use or sell the outcome, it is possible to demonstrate how the outcome will be of economic benefit.

13.3 The role of auditors

All but smaller limited companies are legally required to have their accounts audited by auditors. Shareholders (not directors) appoint the auditors and auditors report to the shareholders. Auditors must be professionally qualified. Auditors are required to provide an independent check of the company's accounting records and to carry out their duties objectively, basing their

Key term

Auditors: those who scrutinise and check accounting records prepared by others.

opinions on the evidence they have been able to verify. Auditors are required to confirm (or not) that:

> accounts comply with the requirements of Companies Acts and accounting standards

> provide a true and fair view of the company's financial position

> the accounts are free from significant errors.

The auditors' report

Auditors are required to confirm in what is called an 'unqualified report' that the accounts:

> comply with the requirements of Companies Acts and accounting standards

> provide a true and fair view of the company's financial position.

A 'qualified report' is provided if the auditors have reservations – perhaps they feel the accounts contain significant errors or the directors' report contains statements with which the auditors do not concur.

True and fair concept

The auditors are required to confirm that this concept has been applied in the financial statements presented to shareholders. Where financial statements are judged to be true and fair it means for example:

> that they are free from bias

> provide information about all a company's assets and liabilities

> provide information in accordance with international accounting standards

> provide a fair assessment of the company's cash flows

> that they have been prepared consistently comparing one year with another year.

Where financial statements are judged to be comparable, reliable, relevant and understandable then they also present a true and fair view.

13.4 The role of directors

Directors are appointed by shareholders and they are answerable to the shareholders. Their main responsibilities include:

> ensuring proper accounting records are maintained and that financial statements are prepared in accordance with legal requirements and international accounting standards

> taking care of the company's resources (sometimes referred to as the stewardship role)

> deciding on accounting policies

> supporting the work of the auditors.

Directors' report

Directors are also required to prepare an annual report to shareholders which outlines how they have discharged their role. As well as information about each director, the report includes details of:

> important events that occurred during the financial year

> shares purchased or sold

X **Common error**

Make sure you don't state that auditors report to the directors (auditors report to the shareholders).

X **Common error**

Be careful not to state that auditors prepare the financial statements (this is the responsibility of the directors).

Key term

Directors: individuals who are responsible for the day-to-day running of a company (executive directors) or providing advice on running a company (non-executive directors).

- ➤ proposed dividends
- ➤ donations to political parties and charities
- ➤ significant future events.

Shareholders' responsibilities

These include electing directors to act on their behalf and to appoint auditors. However, shareholders do not normally have the right to interfere in the day-to-day management of the company.

13.5 What are the greatest challenges?

- ➤ Remembering the number and title of the accounting standards covered by the specification.
- ➤ Preparing statements of cash flow in accordance with IAS7.
- ➤ Applying accounting standards to particular situations.
- ➤ Understanding and distinguishing between the roles of directors, shareholders and auditors.

13.6 Review some important techniques

Preparing a statement of cash flows (in accordance with IAS7)

The statement has a specific layout with precisely worded sections and labels for subtotals, all of which have to be memorised. Here is the format with the wording for subheadings and subtotals emphasised in bold.

Statement of cash flows for the year ended ...

	$000	$000
Operating activities		
Profit from operations	x	
Add depreciation	x	
Add loss on sale/Less profit on sale of non-current assets	x*	
Add decrease in inventories/Less increase in inventories	x*	
Add decrease in trade receivables/Less increase in trade receivables	x*	
Add increase in trade payables/Less decrease in trade payables	x*	
	x*	
Less interest paid during year	(x)	
Less tax paid during year	(x)	
Net cash used in/from operating activities		x*
Investing activities		
Purchase of non-current assets	(x)	
Proceeds from the sale of non-current assets	x	
Net cash used in/from investing activities		x*

Financing activities

Proceeds from the issue of share capital/repayment of shares	x*	
Proceeds from the issue of long-term borrowing/repayment of long-term borrowings	x*	
Dividends paid	(x)	
Net cash used in/from financing activities		x*
Net increase/decrease in cash and cash equivalents		x*
Cash and cash equivalents at beginning of year		x*
Cash and cash equivalents at end of year		x*

*figures which could be negative and which should, therefore, be shown in brackets.

Some of the details require quite complex calculations, so it is important to provide detailed workings.

▼ **Example 13.1** Calculations concerning non-current assets

An extract from the financial statements of R plc is as follows:

Statement of financial position at 31 December

	2016	2015
Assets	$000	$000
Non-current assets	1 960	1 815

Additional information:

During the year ended 31 December 2016 the directors revalued property by $450 000 and disposed of non-current assets with a net book value of $80 000 for $37 000. There were no purchases of non-current assets during the year.

From this limited information it is possible to calculate details required for the cash flow statement.

Depreciation charges

	$000
Original net book value (carrying amount)	1 815
Add revaluation	450
	2 265
Less disposal net book value	(80)
	2 185
Less closing carrying amount	(1 960)
Depreciation for year	225

Disposal profit/loss

	$000
Net book value	80
Disposal proceeds	(37)
Loss on disposal	43

Notes:

➤ The revaluation of property does not appear in the statement of cash flows as it did not generate any cash funds (i.e. it is a non-cash item).

➤ Depreciation charges and the loss on disposal are added back to the operating profit in order to cancel out the effect of the deduction in the income statement. This is because they are also non-cash items.

A bonus issue of shares is a further example of a non-cash item which should not appear in the statement of cash flows.

Preparing a schedule of non-current assets

The format prescribed for a schedule of non-current assets is as follows.

Schedule of non-current assets at ...

	Property $000	Plant $000	Equipment $000
Cost			
At (beginning of year)	x	X	x
Revaluation	x		
Additions		X	x
Disposals			(x)
At (end of year)	x	X	x
Depreciation			
At (beginning of year)	x	X	x
Revaluation	(x)		
Provision for year	x	X	x
Disposals			(x)
At (end of year)	x	X	x
Net book value			
At (end of year)	x	X	x
At (beginning of year)	x	X	x

Applying international accounting standards

When asked to apply an international accounting standard it is good practice to quote the number and title of the standard, to explain the rule being applied and then apply the rule.

▼ Example 13.2 Applying international accounting standards

Situation	Treatment
A limited company owns equipment which cost $36 000 and on which depreciation of $17 000 has been provided. At 31 December 2016 it is considered that the equipment has a fair value of $15 500 and a value in use of $16 200. How should these matters be recorded in the financial statements prepared at this date?	*State the relevant standard* IAS36 Impairment of assets *Explain the rule* The standard requires an impairment to be recorded where the higher of fair value or value in use is less than the carrying amount. *Apply the rule* The carrying amount is $19 000. The higher of fair value ($15 500) and value in use ($16 200) is $16 200. As this is lower than the carrying value, an impairment should be recorded in the income statement of $2 800. The equipment should be recorded in the statement of financial position at the value in use $16 200.
At 31 December 2016, financial statements are being prepared for D Ltd. At this date D Ltd is being sued in the courts following the sale of products to a customer which appear to have had harmful side effects. The directors have been informed that it is possible the case will be lost by the company with estimated costs and damages of $25 000.	*State the relevant standard* IAS37 Provisions, contingent liabilities and contingent assets. *Explain the rule* The standard defines an obligation arising from past events which will lead to a payment to be recorded either as a provision (if probable – more than 50 per cent likely) or as a contingent liability (if possible – less than 50 per cent likely). A provision is shown as a liability in the financial statements with a note giving explanatory details; a contingent liability is not shown in the financial statements, it appears as a note in the accounts unless it is regarded as remote possibility. *Apply the rule* The details should be shown as a contingent liability in the notes to the accounts. This is because the outcome of the court case is possible rather than probable.

Revision checklist

I can:

➤ describe the statutory requirements which apply to limited companies ☐

➤ prepare a schedule of non-current assets ☐

➤ prepare a statement of cash flows ☐

➤ comment on a statement of cash flows ☐

➤ explain the purpose of international accounting standards ☐

➤ identify and outline the main requirements on specific international accounting standards ☐

➤ apply specific international accounting standards ☐

➤ explain the role and responsibilities of auditors and directors ☐

➤ outline key aspects of the true and fair concept. ☐

Raise your grade

The following information is available for P Ltd.

Statement of financial position at 31 December

	2016		2015	
Assets	$000	$000	$000	$000
Non-current assets		474		486
Current assets				
Inventories	49		38	
Trade and other receivables	27		32	
Cash and cash equivalents	11		–	
		87		70
Total assets		561		556
Equity				
Ordinary share capital		250		250
Share premium		60		60
Retained earnings		27		103
Total equity		337		413
Liabilities				
Non-current liabilities				
7% Debentures (2027–28)		200		100
Current liabilities				
Trade and other payables	17		29	
Taxation	7		8	
Cash and cash equivalents	–		6	
		24		43
Total equity and liabilities		561		556

Income statement (Extracts)
for the year ended 31 December

	2016	2015
	$000	$000
Profit from operations	123	118
Finance costs	(14)	(7)
Profit before tax	109	111
Taxation	(7)	(8)
Profit for the year	102	103

Additional information – during the year ended 31 December 2016:

➤ dividends paid totalled $178 000

➤ additional non-current assets were purchased for $36 000

➤ non-current assets with a net book value of $23 000 were sold for $17 000.

Prepare a statement of cash flows for the year ended 31 December 2016.

Student answer

P Ltd.

Statement of cash flows for the year ended 31 December 2016

	$000	$000
①		
Profit from operations	123	
Depreciation ② ③	(25)	
Loss on sale ② ③	(6)	
Less increase in inventories	(11)	
Add decrease in trade receivables	5	
Less decrease in trade payables	(12)	
	74	
Less interest paid during year	(14)	
Less tax ④	(7)	
⑤		53
Financing activities		
Proceeds from the issue of long-term borrowings	100	
Dividends paid	(178)	
⑤		(78)
Investing activities ⑥		
Purchase of non-current assets	(36)	
Proceeds from the sale of non-current assets ⑦	23	
⑤		(13)
Net increase/decrease in cash and cash equivalents		(38)
Cash and cash equivalents at beginning of year		(6)
Cash and cash equivalents at end of year		11

How to improve this answer

① The subheading "Operating activities" has been omitted.

② The depreciation and loss on sale should be added back to the operating, not deducted.

③ Workings for these more complex calculations should be included.

④ The student has selected the current tax provision; the statement should include tax paid which, in this case, is last year's provision.

⑤ Each subtotal should be labelled appropriately.

⑥ The sections for investing activities and financing activities should appear in the reverse order to that shown.

⑦ The proceeds from the sale should be the cash received and not, as here, the net book value of non-current assets.

P Ltd.

Statement of cash flows for the year ended 31 December 2016

	$000	$000
Operating activities		
Profit from operations	123	
Add depreciation **(W1)**	25	
Add loss on sale **(W2)**	6	
Less increase in inventories	(11)	
Add decrease in trade receivables	5	
Less decrease in trade payables	(12)	
	159	
Less interest paid during year	(14)	
Less tax paid during year	(8)	
Net cash used in/from operating activities		114
Investing activities		
Purchase of non-current assets	(36)	
Proceeds from the sale of non-current assets	17	
Net cash used in/from investing activities		(19)
Financing activities		
Proceeds from the issue of long-term borrowings	100	
Dividends paid	(178)	
Net cash used in/from financing activities		(78)
Net increase/decrease in cash and cash equivalents		17
Cash and cash equivalents at beginning of year		(6)
Cash and cash equivalents at end of year		11

W1 Depreciation	$000
Net book value at 1 January	486
Add additions	36
Less disposals at net book value	(23)
	499
Less net book value at 31 December	474
Depreciation	25

W2 Loss on sale of non-current assets	$000
Net book value	23
Less sale proceeds	(17)
loss on sale	6

? Exam-style questions

1 State which international accounting standard should be applied in each of the following situations with regard to a company's published accounts:

(a) Directors of a limited company have discovered that there were material errors in the financial statements.

(b) The board of directors have agreed that the company's property should be revalued.

(c) At the end of the financial year the directors have proposed a dividend on ordinary shares.

2 Explain what is meant by a qualified auditors' report.

3 The following situations have arisen affecting the financial affairs of M Ltd. The company's financial statements are being prepared for the year ended 31 December 2016.

(a) Inventory at 31 December 2016 includes 38 damaged items. These items cost $48 each and would normally be sold for $62 each. It is estimated that 30 items could be repaired at a cost of $14 per item and sold for $52 each; the remaining items will need repackaging at an estimated total cost of $72 and could then be sold for $58 each.

(b) M Ltd. are suing a maintenance company for $40 000 for failing to carry out their contractual duties. The case is before the courts but judgement will not be reached until late January 2017. The directors have been advised that there is 75 per cent chance the company will win its case.

(c) A review of the company's non-current assets at 31 December 2016 has shown that a motor vehicle which cost $45 000 and which has been depreciated by $28 000, has a fair value of $14 000 and a value in use of $13 000.

4 The published accounts of T plc included the following information.

Statements of financial position at 31 December

Assets	2016 $000	2015 $000
Non-current assets	4 188	3 881
Current assets		
Inventories	171	193
Trade and other receivables	97	82
Cash and cash equivalents		31
	268	306
Total assets	4 456	4 187
Equity and reserves		
Share capital $2 ordinary shares	4 000	3 000
Revaluation reserve	-	400
Retained earnings	145	270
	4 145	3 670

Non-current liabilities		
Debentures 8%	220	440
Current liabilities		
Trade and other payables	38	29
Tax liabilities	31	48
Cash and cash equivalents	22	
	91	77
Total equity and liabilities	4456	4187

Additional information:

1   The income statement for the year ended 31 December 2016 included finance costs of $32 000 and a tax provision of $27 000.

2   Dividends paid during the year ended 31 December 2016 were $150 000.

3   The company made a bonus issue of shares of 1 August 2016. No other shares were issued during the year ended 31 December 2016. The issue was made in such a way as to maintain maximum dividend flexibility.

4   Non-current assets: equipment which had a net book value of $180 000 was sold for $193 000 on 1 May 2016; additional plant costing $620 000 was purchased on 1 October 2016.

(a) Prepare a statement of cash flows for the year ended 31 December 2016.

(b) Assess the value of this statement of cash flows to a shareholder in T plc.

14 Business purchase and merger

AL 1.2.1 SB pages 332–353

Key topics

➤ reasons for merging two or more businesses

➤ potential drawbacks arising from a merger

➤ types of merger

➤ business purchase

➤ greatest challenges

➤ important techniques: mergers

➤ important techniques: business purchase.

✓ What you need to know

How to explain the nature and purpose of the merger of different types of business. How to record using ledger accounts: the merger of two or more sole traders to form a partnership; the merger of a sole trader with an existing partnership; the acquisition of a sole trader or partnership by a limited company; the purchase of a business by a limited company. How to prepare statements of financial position post a merger. How to evaluate and discuss the advantages and disadvantages of a proposed merger.

14.1 Reasons for merging two or more businesses

Businesses merge when there is a belief that a combined business will be more successful. The benefits could be of several different kinds:

➤ **Greater market presence**: before a merger one business may be strong in one area but weak in another area, whereas the other business in the merger could be in the reverse position – i.e. strong where the other business is weak; weak where the other business is strong. Where businesses complement each other and are combined they are described as having 'synergy'.

➤ **Economies of scale**: a combined business may be able to operate more efficiently, for example by having joint service departments (such as human resources), and also by gaining economic advantages in production (such as larger trade discounts through making combined orders for materials).

➤ **Vertical integration**: in some mergers a manufacturing organisation merges with its main distributor. As a result the merged business becomes more competitive because the distributor's profit margin is eliminated.

➤ **Research and development**: joint rather than separate research and development activities with the pooling of resources can increase the chance of successful outcomes.

➤ **Diversification**: by forming a larger organisation it may be possible to diversify the range of services or products provided and so increase market share.

➤ **Higher status**: so that as a result of a merger a business could find it easier to attract finance or a more skilled workforce.

14.2 Potential drawbacks arising from a merger

➤ **High costs of merger**: can arise as duplicated resources are scaled down with the disposal of some assets, including property, often at a substantial loss. Staff redundancies are also highly probable leading to significant redundancy payments.

➤ **Loss of staff morale**: can arise partly because many individuals will feel threatened by the redundancies which are likely to take place, but also because they are unsettled by the change of job roles in the new organisation.

14.3 Types of merger

➤ **Two or more sole traders merge to form a partnership**: the individuals concerned will decide exactly which assets and liabilities will be transferred to the new business, maybe retaining some assets for private use or disposing of them prior to the merge. The capitals of each individual in the new business will depend on the outcome of these decisions. The partners in the new business must decide how they will share profits and losses.

➤ **Two or more sole traders form a limited company**: as a result the individuals concerned gain the significant advantage of limited liability. Each individual involved in the merger will be given shares in the new business in proportion to the value of the net assets they provide. Since each share carryies voting rights, it follows that control in the limited company will be proportionate to each individual's investment.

➤ **The merge of a partnership with a sole trader**: the individuals concerned will agree on what assets and liabilities will be transferred to the new business, and a new partnership agreement will be required.

14.4 Business purchase

When a limited company purchases another business it is possible for it to settle the purchase price in a number of ways:

➤ cash

➤ shares (at face value or at a premium) which are allocated to the owner (or owners) of the business being purchased

➤ debentures which are allocated to the owner (or owners) of the business being purchased

➤ a combination of any of cash, shares, debentures.

14.5 What are the greatest challenges?

➤ Absorbing a considerable amount of detail and ensuring that all relevant matters are taken into account in the answer.

➤ Correctly calculating the goodwill or negative goodwill when a limited company purchases another business.

➤ Avoiding confusing details relevant to a purchase and details relevant to a sale in questions where both aspects are covered.

➤ Making accurate calculations of the value of shares and the number of shares issued by a limited company in business purchases and mergers.

➤ Preparing ledger records for a purchase or for a sale.

> **Key terms**
>
> **Negative goodwill:** when the amount paid to acquire a business's net assets is less than their net book value.

14.6 Review some important techniques: mergers

Two sole traders merge to form a partnership

▼ **Example 14.1** R and S have agreed to merge their two businesses to form a partnership. Prior to the merger their summarised statements of financial position were as follows:

	R	S
	$	$
Non-current assets	85000	97000
Current assets	32000	21000
Total assets	117000	108000
Capital	72000	89000
Non-current liabilities	30000	–
Current liabilities	15000	19000
Capital and liabilities	117000	108000

It has been agreed that:

➤ R will retain for private use his business's motor vehicle which has a net book value of $16000.

➤ R will retain for private use his business's inventory which has a value of $8000.

➤ R will take responsibility for paying off his business's loan and S will take responsibility of paying off his trade payables value $5000 and these will not be brought into the partnership.

➤ R's remaining non-current assets will be revalued at $90000 and S's non-current assets will be revalued at $115000.

The summarised statement of financial position of the new business immediately after the merger will be as follows:

	Workings		$
	from R	from S	
Non-current assets	$90000	$115000	205000
Current assets	£32000 – $8000 inventory = $24000	$21000	45000
		Total assets	250000
Capital R	A $114000 – L 15000 = $99000		99000
Capital S		A $136000 – L $14000	122000
Current liabilities	$15000	$19000 – $5000 = $14000	29000
		Capital and liabilities	250000

Two sole traders merge to form a limited company

> ▼ **Example 14.2** Using the information in Illustration 1, but assume that R and S agree to form a limited company and that each will receive $1 of ordinary shares for each $1 invested in the new business.
>
> The summarised statement of position will be similar to that shown for Illustration 1, except that the partners' capitals will be replaced by the issued share capital of the company:
>
> | | $ | |
> |---|---:|---|
> | Non-current assets | 205 000 | |
> | Current assets | 45 000 | |
> | Total assets | 250 000 | |
> | Equity | | |
> | Ordinary shares | 221 000 | R will own 99 000 $1 shares; |
> | | | T will own $122 000 $1 shares |
> | Current liabilities | 29 000 | |
> | Equity and liabilities | 250 000 | |

A partnership merges with a sole trader (including goodwill)

> ▼ **Example 14.3** It has been agreed that the partnership of J and K will merge with the partnership of L. Prior to the merger the businesses' summarised statements of financial position were as follows:
>
> | | J and K | L |
> |---|---:|---:|
> | | $ | $ |
> | Non-current assets | 149 000 | 49 000 |
> | Current assets | 37 000 | 9 000 |
> | Total assets | 186 000 | 58 000 |
> | Capital: J | 100 000 | |
> | Capital: K | 60 000 | |
> | Capital: L | | 45 000 |
> | Non-current liabilities | 10 000 | 5 000 |
> | Current liabilities | 16 000 | 8 000 |
> | Capital and liabilities | 186 000 | 58 000 |
>
> J and K were equal partners. In the new business profits and losses will be shared in the ratio J:K:L, 4:3:3.
>
> It has been agreed that:
>
> ➤ the goodwill of the partnership will be valued at $50 000 and that of L's business at $30 000
>
> ➤ goodwill will not be recorded in the books of the new business
>
> ➤ non-current assets with a net book value of $22 000 will be retained by the J for private use and will not be transferred to the new business

- ➤ remaining tangible non-current assets of the partnership will be valued at $145 000

- ➤ L's non-current assets will be transferred to the new business at a value of $65 000

- ➤ the current assets and current liabilities of the two businesses will be transferred to the new business at their book values

- ➤ K has agreed to settle the partnership's non-current liabilities privately

- ➤ L will settle his business's non-current liability privately

- ➤ the new business will continue to use the books of the partnership and will be known as JKL.

Step 1: calculating the capital of L in the new business (before writing off goodwill)

	$
Goodwill	30 000
Tangible non-current assets (at valuation)	65 000
Current assets	9 000
Current liabilities	(8 000)
	96 000

Step 2: calculating the capitals of J and K in the new business (before writing off goodwill)

	Workings and notes	J	K
		$	$
Original capitals		100 000	60 000
Non-current assets retained		(22 000)	
Revaluation surplus	NCA revalued at $145 000 – book value of NCA ($149 000 – $22 000), i.e. surplus of $18 000 shared equally	9 000	9 000
Goodwill	$50 000 shared equally	25 000	25 000
Settlement of non-current liability	K settles the non-current liability of $10 000 privately, this is the equivalent of making an additional capital contribution		10 000
Capitals in new business		112 000	104 000

Step 3: writing off the goodwill

The goodwill to be written off totals: $50 000 + $30 000 = $80 000

	J	K	L
	$	$	$
Capitals from Steps 1 and 2	112 000	104 000	96 000
Goodwill written off in ratio 4:3:3	(32 000)	(24 000)	(24 000)
Capitals in new business	80 000	80 000	72 000

> **Remember**
>
> To use the ratio agreed for the new business to write off the goodwill.

Step 4: preparing the (summarised) statement of financial position

JKL		
Statement of financial position		
	$	**Workings**
Non-current assets	210 000	From JK $145 000; from L $65 000
Current assets	46 000	From JK $37 000; from L $9 000
Total assets	256 000	
Capital accounts		
J	80 000	
K	80 000	
L	72 000	
Total capitals	232 000	
Current liabilities	24 000	From JK $16 000; from L $8 000
Total capital and liabilities	256 000	

14.7 Review some important techniques: business purchase

▼ **Example 14.4** A limited company purchasing a sole trader; purchase consideration settled in shares

The directors of F Limited has agreed to purchase the sole trader business owned by Jing. The purchase consideration will be settled by the issue of 140 000 shares of 50 cents each issued at a premium of 20 per cent.

The summarised statements of financial position of the two businesses immediately before the merger were as follows:

	F Limited	Jing
	$	$
Non-current assets	650 000	48 000
Current assets	64 000	9 000
Cash and cash equivalents	16 000	8 000
	730 000	65 000
Share capital/capital	520 000	58 000
Retained earnings	170 000	
Current liabilities	40 000	7 000
	730 000	65 000

It has been agreed that Jing's non-current assets will be revalued at $60 000 and current assets (excluding cash and cash equivalents) will be revalued at $6000. Jing will use his business's cash resources to discharge his business's current liabilities.

Step 1: Calculate the purchase consideration: 140 000 shares of 50 cents each = $70 000 plus share premium of 20 per cent (20% of $70 000) $14 000 = $84 000

> **Key term**
>
> Purchase consideration: the agreed amount to be paid to acquire a business.

> **Remember**
>
> Keep separate in your mind the business purchase from the purchaser's point of view and the sale of the business from the vendor's point of view.

Step 2: Consider the details from the point of view of the purchaser
(F Limited)

The company will pay $80 000 and acquire assets valued at:

	$
Non-current assets	60 000
Current assets	6 000
	66 000

So the goodwill arising from the purchase is $18 000 (purchase consideration $84 000 less value of assets taken over $66 000).

> **Remember**
>
> In a business purchase the extra amount paid for the (net) assets purchased is goodwill.

F's Limited's statement of financial position after the purchase will show:

	$	Workings
Goodwill	18 000	
Tangible non-current assets	710 000	Original $650 000 + $60 000
Current assets	70 000	Original $64 000 + $6 000
Cash and cash equivalents	16 000	
	814 000	
Share capital	590 000	Original $520 000 + new $70 000
Share premium	14 000	New $14 000
Retained earnings	170 000	
Current liabilities	40 000	
	814 000	

> **Remember**
>
> Purchased goodwill can appear on a limited company's statement of financial position as in this illustration.

Step 3: Consider the details from the point of view of the vendor (Jing)

Jing will sell his non-current assets and current assets (excluding cash and cash equivalents) and receive shares valued at $84 000.

Jing will make a profit on the sale of his business:

	$	$
Value of shares received from F Limited		84 000
Less: Non-current assets sold	48 000	
Current assets sold	9 000	
		(57 000)
Profit on sale of business		17 000

> **Remember**
>
> In the sale of a business, a profit is made if the sale price exceeds the value of the (net) assets sold to the purchaser.

Jing will use the business's cash resources ($8000) to settle his business's current liabilities of $7000. So he will have $1000 cash and cash equivalents to keep for himself.

Summary of Jing's statement of financial position immediately after the sale of his business:

	$
Assets	
Shares in F Limited	84 000
Cash and cash equivalents	1 000
Total assets	85 000
Capital	
Just before sale to F Limited	58 000
Add profit on sale of business	17 000
Total capital	85 000

▼ **Example 14.5** A limited company purchasing a partnership; purchase consideration settled in shares, debentures and cash; ledger records of a purchase and of a sale

Z Limited has arranged to purchase the partnership of Jay and Kay. Jay and Kay share profits and losses in the ratio Jay:Kay, 3:2.

Z Limited have agreed to take over all the partnership's assets (except cash and cash equivalents) and current liabilities for a purchase consideration of $180 000. The purchase consideration will be discharged by the issue of 90 000 $1 ordinary shares in Z limited at a premium of 40 cents per share, $50 000 of 6 per cent debentures and the balance in cash. The company has decided to revalue the partnership's non-current assets at $140 000 and the current assets which it takes over at $25 000.

The statements of financial position of the two businesses prior to the purchase of the partnership by Z Limited were as follows:

	Z Limited	Jay and Kay
	$	$
Non-current assets	820 000	113 000
Current assets	49 000	30 000
Cash and cash equivalents	16 000	12 000
	885 000	155 000
Share capital	830 000	
Capital Jay		90 000
Capital Kay		50 000
Current liabilities	55 000	15 000
	885 000	155 000

Step 1: consider the details from the point of view of Z Limited

Calculation of goodwill involved in purchase consideration

The company will pay $180 000 and acquire net assets valued at:

	$
Non-current assets	140000
Current assets	25000
	165000
Less current liabilities	15000
Net value	150000

So the goodwill arising from the purchase is $30 000 (purchase consideration $180 000 less value of net assets taken over $150 000).

The purchase consideration will be settled as follows:

	$
Shares: 90 000 $1 shares at premium of 40 cents per share (i.e. 90 000 x $1.40)	126000
6% Debentures	50000
Balance in cash	4000
Total: purchase consideration	180000

In Z Limited's ledger it will be necessary to open a business purchase account:

Business purchase account

	$		$
Current liabilities	15000	Non-current assets	140000
Purchase consideration	180000	Current asses	25000
		Goodwill	30000
	195000		195000

An account will also be required to show the amount due to the partnership and how this amount is settled:

Partnership of Jay and Kay

	$		$
Shares	126000	Purchase consideration	180000
Debentures	50000		
Cash	4000		
	180000		180000

Z Limited's statement of financial position immediately after the purchase of the partnership is as follows:

		Workings
	$	
Goodwill	30000	
Tangible non-current assets	960000	Original $820 000 + $140 000
Current assets	74000	Original $49 000 + $25 000
Cash and cash equivalents	12000	Original $16 000 − cash paid $4 000
	1076000	

Exam tip

Sometimes the business purchases account is named after the vendors of the business, e.g. in the illustration the account would be called 'Jay and Kay account', so an exam question could ask for either of these.

Key term

Vendors: the owner or owners of the business which is being sold.

Share capital	920 000	Original $830 000 + new $90 000
Share premium	36 000	New 90 000 × 40 cents
6% Debentures	50 000	
Current liabilities	70 000	Original $55 000 + $15 000
	1 076 000	

> ★ **Exam tip**
>
> When considering the partnership's situation the records required are the same as for a dissolution of partnership (see Unit 6).

Step 2: consider the details from the point of view of the partnership of Jay and Kay

Calculate the profit on sale of the business to Z Limited:

As a calculation			As a ledger account			
	$	$	**Realisation account**			
				$		$
Sale price		180 000	Non-current assets	113 000	Current liabilities	15 000
Less value of net assets sold to company:			Current assets	30 000	Z Limited: sale price	180 000
Non-current assets	113 000					
Current assets	30 000		Profit: Jay	31 200		
	143 000		Profit: Kay	20 800		
Less current liabilities	(15 000)			195 000		195 000
		(128 000)				
Profit on sale		52 000				
The profit of $52 000 is shared between the partners: Jay (3/5ths) $31 200; Jay (2/5ths) $20 800						

Record how the partners will share out the proceeds of the sale to the limited company and close the business.

Option 1: the partners agree to divide the shares and debentures in their profit sharing ratio

As a calculation:

	Jay	Kay
	$	$
Original capital balances	90 000	50 000
Profit on sale of business	31 200	20 800
	121 200	70 800
Less division of shares (90 000 $1 shares value $126 000)		
So Jay receives 3/5 × 90 000 shares, value 3/5 × $126 000	(75 600)	
And Kay receives 2/5 × 90 000 shares, value 2/5 × $126 000		(50 400)
Less division of 6% Debentures value $50 000 in ratio 3:2	(30 000)	(20 000)
Amount still outstanding	15 600	400
Share of cash paid out to settle the amounts due		
Cash available: original balance $12 000 + $4 000 from Z Limited, i.e. $16 000	(15 600)	(400)

As ledger accounts:

Capital accounts

	Jay	Kay		Jay	Kay
	$	$		$	$
Shares	75 600	50 400	Opening bal	90 000	50 000
Debentures	30 000	20 000	Profit on sale	31 200	20 800
Bank	15 600	400			
	121 200	70 800		121 200	70 800

Bank account

	$		$
Opening balance	12 000	Capital Jay	15 600
Z Limited	4 000	Capital Kay	400
	16 000		16 000

Z Limited account

	$		$
Agree sale price	180 000	Shares	126 000
		Debentures	50 000
		Capital Kay	4 000
	180 000		180 000

Option 2: the partners decide to divide the shares and debentures equally between them.

	Jay	Kay
	$	$
Original capital balances	90 000	50 000
Profit on sale of business	31 200	20 800
	121 200	70 800
Less equal division of shares	(63 000)	(63 000)
Less division of 6% Debentures value $50 000	(25 000)	(25 000)
Balance	33 200	(17 200)

In this situation Jay is still owed $33 200, but Kay is over-rewarded and owes the partnership $17 200.

So Jay receives all the remaining cash $16 000 and Kay will need to make a payment to Jay of $17 200 from private resources to settle the balance due to Jay.

I can:

➤ prepare records to record the merger of two sole traders and the merger of a sole trader and a partnership ☐

➤ calculate goodwill involved in the acquisition of a business ☐

➤ prepare records for the acquisition of a sole trader or a partnership by a limited company ☐

➤ prepare a business purchase account ☐

➤ prepare records to dissolve a partnership when it is acquired by a limited company ☐

➤ prepare financial statements immediately after the merger of two or more businesses ☐

➤ prepare financial statements after a limited company has acquired another business ☐

➤ comment on the potential benefits and disadvantages of a proposed merger or acquisition. ☐

 Raise your grade

B Limited has been formed to take over the partnership of Bikash and Rakesh. The partners share profits and losses in the ratio Bikash:Rakesh, 4:3. At the date of the take over the partnership's statement of financial position was as follows:

	$	$
Non-current assets (net book value)		88 000
Current assets		
Inventory	22 000	
Trade receivables	14 000	
Cash and cash equivalents	17 600	
		53 600
Total assets		141 600
Capital accounts		
Bikash	80 000	
Rakesh	40 000	
		120 000
Loan from Rakesh (8% interest per annum)		12 000
Current liabilities		
Trade payables		9 600
Total capital and liabilities		141 600

All the assets (except cash and cash equivalents) and current liabilities of the partnership are to be taken over by B Limited. The assets are to be valued as follows:

	$
Non-current assets	125 000
Inventory	17 000
Trade receivables	12 000

The purchase consideration is to be $160 000 satisfied as follows:

- 90 000 ordinary shares of $1 each will be issued at a premium of 25 cents per share and will be divided among the partners in proportion to their capitals.

- Rakesh will be given sufficient 6 per cent debentures to ensure he receives the same amount of interest per annum as he received on his loan to the partnership.

- Any balances remaining on the partners' capital accounts to be settled in cash through B Limited's bank account.

(a) Calculate:

- the goodwill arising on the acquisition of the partnership

- the profit or loss made by the partnership on its dissolution.

(b) Prepare Rakesh's capital account recording the closing of the partnership's books.

Student answer

(a)

Goodwill calculation	$	$
Purchase consideration		160 000
Non-current assets	125 000	
Inventory	17 000	
Trade receivables ❶	14 000	
Cash ❷	17 600	
	173 600	
less liabilities ❸	(21 600)	
		(152 000)
Goodwill		8 000

Profit on dissolution	$	$
Purchase consideration		160 000
Non-current assets	125 000	
Inventory ❶	22 000	
Trade receivables	14 000	
	161 000	
Current liabilities	(9 600)	
		(151 400)
Profit on dissolution		8 600

(b)

BOOKS OF BIKASH AND RAKESH

Capital account (Rakesh)

Dr			Cr	
	$			$
Shares ❺	45 000	Opening balance		40 000
Debentures ❹	12 000	Profit on dissolution ❻		3 440
		Cash ❼		13 560
	57 000			57 000

How to improve this answer

1. There has been confusion about which figures to use in the calculations: when looking at the details from the purchaser's point of view the assets at revaluation should be used; when looking at the information from the vendors' point of view the net book values in the partnership's books of account should be used.

2. Cash is not being taken over so should not be included in the calculation.

3. Only current liabilities are being taken over so, so the loan should not have been included in the calculation.

4. The instruction relating to the value of debentures to be issued has been ignored and no workings have been provided.

5. The candidate has divided the shares in the profit sharing ratio and not, as requested, in proportion to the value of the partners' capitals; no workings have been provided.

6. The profit on dissolution has been divided correctly, but there were errors in calculating the profit in task (a).

7. As a result of other errors, the candidate has assumed that the partner must pay in cash from his private funds.

Model answer

(a)

Goodwill calculation	$	$
Purchase consideration		160 000
Non-current assets	125 000	
Inventory	17 000	
Trade receivables	12 000	
	154 000	
less current liabilities	(9 600)	
		(144 400)
Goodwill		15 600

Profit on dissolution	$	$
Purchase consideration		160 000
Non-current assets	88 000	
Inventory	22 000	
Trade receivables	14 000	
	124 000	
less current liabilities	(9 600)	
		(114 400)
Profit on dissolution		45 600

(b)

W1: Debentures – must provide same interest as loan, i.e. 8% × $12 000 = $960; so debentures issued × 6/100 = $960, i.e. $16 000 debentures

W2: Profit on dissolution – total profit $45 600 × 2/5 = $18 240

W3: Shares – 1/3 × ($90 000 shares + $22 500 share premium) i.e. $37 500

BOOKS OF BIKASH AND RAKESH

Capital account (Rakesh)

Dr			Cr
	$		$
Shares (**W3**)	37 500	Opening balance	40 000
Debentures (**W1**)	16 000	Loan	12 000
Cash	16 740	Profit on dissolution (**W2**)	18 240
	70 240		70 240

Key features of the model answer

➤ When recording details in the purchaser's books of account, the valuations used are those the purchaser has placed on the assets which have been taken over.

➤ When recording details in the vendors' books of account, the valuations used for assets are those recorded in the partnership's statement of financial position.

➤ The partnership cash and loan have been excluded from the calculations of goodwill and profit on dissolution.

➤ The business purchase account records the details of the assets and liabilities that are being acquired and two separate debit entries are made for shares (one to match the credit entry which would be made in the share capital account, and a second entry to match the credit entry which would be made in the share premium account).

➤ Workings have been provided for all complex calculations.

➤ The correct ratio has been used to divide the shares.

➤ The figure for debentures to be given to Rakesh has been calculated in line with the requirements of the question.

? Exam-style questions

1 State **three** ways in which a limited company could settle a purchase consideration when acquiring another business.

2 Describe **three** reasons why two businesses might agree to merge.

3 Explain how goodwill is calculated when a business purchases another business.

4 Mary and Nathan are in partnership sharing profits and losses equally. They have decided to merge their business with the business owned by Oliver. It was agreed that in the new business profits and losses will be shared: Mary, 40 per cent, Nathan 40 per cent, Oliver 20 per cent.

The statements of financial position for each business immediately before the merger were as follows.

	Mary and Nathan	Oliver
	$	$
Non-current assets	245 000	82 000
Current assets		
Inventory	17 000	9 000
Trade receivables	14 000	6 000
Cash and cash equivalents		7 000
Total assets	276 000	104 000

Capital: Mary	130 000	
Capital: Nathan	115 000	
Capital: Oliver		95 000
Current account: Mary	(3 000)	
Current account: Nathan	5 000	
Non-current liabilities		
Bank loan	11 000	
Current liabilities		
Trade payables	10 000	9 000
Bank overdraft	8 000	
Capital and liabilities	276 000	104 000

It was agreed that:

- The goodwill of each business will be valued at $60 000 (Mary and Nathan) and $40 000 (Oliver); a goodwill account will not be maintained in the books of the new business.

- Oliver will retain his business's motor vehicle, net book value $18 000, for private use, and he will settle his business's current liabilities from private funds; all other assets will be transferred to the new business at their net book value.

- All of the partners' assets will be transferred to the new business, but non-current assets will be revalued at $290 000 and inventory $14 000.

- Mary will settle the bank loan from her private resources.
- The balances of the partners' current accounts will be transferred to their capital accounts.
- The books of the partnership would continue to be used in the new business which would be called MNO.

Prepare:

(a) The capital accounts of the three partners in the books of the partnership to show the entries arising from the merger.

(b) A statement of financial position of MNO immediately after completion of the merger.

5 E Limited has arranged to purchase the partnership of Kerry and Larry who have been sharing profits and losses equally.

E Limited have agreed to take over all the partnership's non-current assets, inventory and current liabilities for a purchase consideration of $170 000. The purchase consideration will be discharged by the issue of 10 per cent debentures to Kerry to ensure that she receives the same amount of interest per annum as was the case with her loan to the partnership. The balance of the amount due will be settled by issuing ordinary shares of $1 each at a premium of 25 cents per share. The company has decided to revalue the partnership's non-current assets at $145 000 and inventory at $18 000.

The statements of financial position of the two businesses prior to the purchase of the partnership by E Limited were as follows:

	E Limited	Kerry and Larry
	$	$
Non-current assets	635 000	116 000
Inventory	36 000	24 000
Trade receivables	23 000	8 000
Cash and cash equivalents	25 000	17 000
	719 000	165 000
Share capital	600 000	
Retained earnings	85 000	
Capital Kerry		75 000
Capital Larry		50 000
Loan (12.5%) Kerry		16 000
Current liabilities	34 000	24 000
	719 000	165 000

The partnership's trade receivables settled their accounts in full.

Prepare:

(a) E Limited's statement of financial position immediately after the acquisition of the partnership.

(b) Calculate the profit made by the partnership on its sale to E Limited.

(c) Prepare the capital accounts of the partners recording the closure of the books of account.

(d) Discuss the impact of the acquisition on the original shareholders of E Limited.

Key topics

> ➤ consignment accounts
>
> ➤ joint ventures
>
> ➤ computerised accounting records
>
> ➤ greatest challenges
>
> ➤ important techniques.

✓ What you need to know

How to prepare consignment accounts and value of unsold inventory. Calculating profit and preparing ledger accounts for joint ventures. The advantages, and disadvantages, of computerised accounting and the process of computerising business accounts.

15.1 Consignment accounts

Consignment accounts are required when a business chooses to sell goods overseas through an agent, benefiting from the agent's local knowledge and contacts and avoiding the cost of establishing its own local trading unit.

When selling goods on consignment the following steps usually occur

Step 1: the consignor sends goods to the consignee for sale. Details of the goods sent are recorded on a pro-forma invoice.

Step 2: the consignee endeavours to sell the goods received and sends an 'account sales' detailing the goods sold, their sales value, and the amount due to the consignor less expenses paid by the consignee.

Consignee's commission

the consignee receives a commission based on the sale proceeds. The commission is included in the expenses deducted from the sale proceeds. There is also the possibility that the consignee will receive an additional commission for taking responsibility for any irrecoverable debts arising from the sales of goods on consignment. Where this arrangement is made the account sales does not shown any deduction for irrecoverable debts; instead this 'del credere commission' is deducted from the amount due to the consignor.

Step 3: the consignee settles the amount due to the consignor.

💡 Remember

The consignor remains the owner of the goods until they are sold by the consignee.

Key terms

Consignor: the business selling goods through an agent.

Consignee: the agent selling goods received from a consignor.

Pro-forma invoice: a document detailing the goods sent but, unlike a normal invoice, not charging the consignee for the goods sent.

Account sales: document sent by a consignee to the consignor recording details of sales, expenses incurred and the net amount due to the consignor.

Del credere commission: a commission charged by a consignee in return for taking responsibility for any irrecoverable debts arising from the sale of goods on consignment.

Accounting records

In the consignor's accounting system:

➤ Consignment outwards account: recording the value of goods sent on consignment.

➤ Consignment to (consignee) account: recording the income and expenses arising from the consignment and the profit (or loss) made.

➤ Personal account for the consignee to record the amounts due from the consignee and payments made in settlement.

In the consignee's accounting system:

➤ Personal account for the consignor: to record the amounts due to the consignor and payments made in settlement.

➤ Commission account(s): to record the consignee's income arising from the consignment.

Valuing unsold inventory

Unsold inventory is valued to take account of the value of unsold goods plus an appropriate proportion of the expenses relating to the unsold consignment.

> **Remember**
>
> When valuing unsold inventory do not include any portion of the commission charges as this expense applies only to the goods sold.

15.2 Joint ventures

A joint venture occurs when two or more businesses or individuals work together on a temporary basis on a project or series of projects. Assuming the joint venture is relatively small scale, each participant will keep their own records of relevant transactions. The participants will also agree on how the profit or loss on the joint venture should be divided.

The accounting records required in each participant's books of account are:

Joint venture account to record:

➤ all payments made by that participant (expenses, payments to other participants)

➤ all amounts received by that participant (sale proceeds, amounts received from other participants).

Memorandum joint venture account (which is not part of the double-entry) which is used to calculate the profit or loss on the joint venture and includes:

➤ all expenses of the joint venture (irrespective of who paid the expenses)

➤ all income from the joint venture (irrespective of who received the income).

> **Key term**
>
> **Joint venture**: a short-lived arrangement by two or more participants who act together in the hope of making a profit on a specific project or projects.

15.3 Computerised accounting records

It is now commonplace for businesses to use accounting software packages rather than maintain manual accounting records. The key features of software packages are:

➤ Automatic processing of data: so that the operator keys in the important data from a source document and all accounting records are updated automatically: ledger accounts, trial balances, income statements and statements of financial positions.

➤ Integration of functions: most software packages will also produce other important records which may not be part of a manual system: inventory records, payroll, invoices, credit notes, statements of account, etc.

➤ Provision of management information including: analysis of trade receivable accounts to help identify late payers, audit trails (which track the origins of all figures in the accounting system) ratio analysis reports, break-even analysis, costing schedules, budgets, flexible budgets, investment appraisals, etc.

The main benefits of switching from a manual to a computerised system are:

➤ greater accuracy

➤ greater speed

➤ simultaneous updating of records

➤ improved accessibility

➤ provision of more information for decision making

➤ possibility of reducing levels of staffing.

Possible drawbacks include:

➤ substantial additional costs of equipment, software (especially if tailor-made for the organisation), maintenance, upgrades

➤ additional expenditure on staff training

➤ risk of data loss.

The introduction of a new system usually involves the following:

➤ Decision about whether to buy a standard software package, or invest in something especially designed for the organisation. The decision is often a difficult one to make, not only because of the cost but also because of the difficulty of foreseeing the needs of the organisation a few years ahead.

➤ Transition period where usually the manual system is maintained alongside the new computerised system for an agreed period to ensure that the new system works well and there is no interruption to the smooth running of the organisation.

Security measures to protect data to include:

➤ back-up procedures to ensure that data is not lost through accidental or malicious data loss-up procedures can be made automatic (usually only the most recent inputs are lost where there are effective back-up procedures)

➤ access restrictions: to ensure that only designated members of staff can access sensitive information including the use of passwords

➤ virus protection: introduction of programmes which offer protection

➤ encryption of data to encode sensitive data and so ensure confidentiality

➤ extranet which is a private network which can only be accessed by authorised users.

Those responsible for introducing a computerised system must also be aware of legal requirements particularly regarding data protection.

15.4 What are the greatest challenges?

➤ Consignment accounts: ensuring that transactions are correctly recorded without confusing the entries to be made in the books of the consignor and the books of the consignee.

➤ Consignment accounts: calculating the value of unsold inventory.

➤ Joint ventures: keeping a clear distinction between the records of each participant.

> **X Common error**
>
> Don't assume that computer records will be error free as long as the correct data has been inputted – occasionally faults occur in software programmes which can lead to serious inaccuracies.

15.5 Review some important techniques

▼ **Example 15.1** Ledger accounts for consignments

R Limited sent a consignment of 1500 microwave ovens to its agent, Chatri, in Bangkok in June 2016. The ovens cost $80 each.

The following information is available:

	$
Insurance charges paid by R Limited	4 200
Freight charges paid by R Limited	8 250
Import duties paid by Chatri	18 000
Storage costs paid by Chatri	4 250
Irrecoverable debts	500

On 31 December 2016 an account sales submitted by Chatri recorded the sale of 1200 microwave ovens at $125 each. Chatri is entitled to a commission of 5 per cent on all sales plus an additional del credere commission of 2 per cent on all sales. Chatri remitted $98 000 to R Limited on 31 December 2016.

R Limited's ledger accounts

Consignment outwards

Dr			Cr
	$		$
		Chatri	120 000

Consignment to Chatri, Bangkok

Dr			Cr
	$		$
Consignment outwards	120 000	Chatri, sale proceeds	150 000
Bank: insurance	4 200	Balance c/d (unsold inventory)	
Bank: freight charges	8 250	**(W1)**	30 940
Chatri: import duties	18 000		
Chatri: storage costs	4 250		
Chatri: commission	7 500		
Chatri: del credere commission	3 000		
Profit on consignment	15 740		
	180 940		180 940
Balance b/d (inventory)	30 940		

Chatri, Bangkok

Dr	$	Cr	$
Consignment, sale proceeds	150 000	Consignment expenses:	
		Import duties	18 000
		Storage costs	4 250
		Commission (5% × $150 000)	7 500
		Del credere commission	
		(2% × $150 000)	3 000
		Bank: proceeds	98 000
		Balance c/d	19 250
	150 000		150 000
Balance b/d	19 250		

W1 Closing inventory	$	$
300 items @ $80 each (i.e. 1/5th of consignment)		24 000
Less Proportion of relevant expenses		
insurance	4 200	
freight charges	8 250	
import duties	18 000	
storage costs	4 250	
	34 700	
1/5th of expenses on all goods in consignment		6 940
Value of closing inventory		30 940

Chatri's ledger accounts

R Limited

	$		$
Bank, import duties	18 000	Bank, sale proceeds	150 000
Bank, storage costs	4 250		
Commission	7 500		
Del credere commission	3 000		
Bank: paid on account	98 000		
Balance c/d	19 250		
	150 000		150 000
		Balance b/d	19 250

Commission

	$		$
		R Limited	7 500

Del credere commission

	$		$
		R Limited	3 000

X Common error

Make sure you don't record expenses paid by R Limited in the consignee's accounts.

Remember

The balance of the two commission accounts will be transferred to the consignee's income statement at the year-end; the irrecoverable debts will also appear in the consignee's income statement to offset the income from the del credere commission.

Revision checklist

I can:

➤ consignment accounts: prepare the records of the consignor ☐

➤ consignment accounts: prepare the records of the consignee ☐

➤ consignment accounts: value unsold inventory ☐

➤ joint ventures: prepare ledger accounts and calculate profit on a joint venture ☐

➤ computerised accounts: explain the advantages and disadvantages and the process of computerising business accounts. ☐

Raise your grade

Shafiq and Peishan have agreed to form a joint venture to sell a quantity of office furniture they purchased recently when a business closed down. They have agreed that the profits and losses of the joint venture should be shared equally.

The following details are available.

- Shafiq paid for the office furniture $14 300.

- Peishan paid $640 to collect the office furniture and $820 in storage costs

- It was agreed that Peishan should be paid a wage of $520 for delivery items of office furniture to customers and a commission of 2 per cent of all sale proceeds.

- Shafiq paid advertising costs of $420.

- Peishan sold nine-tenths of the office furniture for $20 500 and paid half to Shafiq.

- Shafiq agreed to keep the unsold furniture for personal use at an agree valuation of $950.

- Final settlement of any amounts due were made between the participants.

Prepare the joint venture accounts in the books of (a) Shafiq and (b) Peishan and (c) a memorandum joint venture account.

Student answer

BOOKS OF SHAFIQ

Joint Venture with Peishan

	$		$
Bank, office furniture	14 300	Bank, receipt from Peishan ❹	17 835
Bank, advertising	420		
Office furniture retained ❶	1 430		
Profit on joint venture	1 685		
	17 835		17 835

BOOKS OF PEISHAN

Joint Venture with Shafiq

	$		$
Bank, collecting furniture	640	Bank, sale proceeds	20 500
Bank, storage costs	820	Wage ❷	520
Profit on joint venture	1 685	Commission ❷ ❸	410
Bank, payment to Shafiq ❹	18 285		
	21 430		21 430

Memorandum Joint Venture Account

	$		$
Office furniture	14 300	Sale proceeds	20 500
Advertising	420		
Office furniture retained ❶	950		
Collection charges	640		
Storage costs	820		
Profit ❷ ❺	3 370		
	20 500		20 500

How to improve this answer

❶ Incorrect treatment of office furniture retained by Shafiq and incorrect valuation used.

❷ Incorrect treatment of allowances made to Peishan for wage and for commission.

❸ No workings provided for the figure for commission.

❹ The payment made by Peishan to Shafiq midway through the joint venture has been ignored; the final payment between participants should be of the same amount if all entries have been made correctly.

❺ The profit figure is incorrect because of other errors but has been correctly divided between the participants.

Model answer

BOOKS OF SHAFIQ

Joint Venture with Peishan

	$		$
Bank, office furniture	14300	Bank, receipt from Peishan	10250
Bank, advertising	420	Office furniture retained	950
Profit on joint venture	2170	Bank, receipt from Peishan	5690
	16890		16890

BOOKS OF PEISHAN

Joint Venture with Shafiq

	$		$
Bank, collecting furniture	640	Bank, sale proceeds	20500
Bank, storage costs	820		
Wage	520		
Commission (**W1**)	410		
Bank, payment to Shafiq	10250		
Profit on joint venture	2170		
Bank, payment to Shafiq	5690		
	20500		20500

W1 Commission: 2% × sale proceeds $20500 = $410

Memorandum Joint Venture Account

Office furniture	14300	Sale proceeds	20500
Advertising	420	Office furniture retained	950
Collection charges	640		
Storage costs	820		
Wage	520		
Commission	410		
Profit	4340		
	21450		21450

★ Exam tips

Key features of the model answer

➤ If a participant retains any assets in a joint venture the item (at an agreed valuation) should be regarded as income and credited to the relevant joint venture account.

➤ Any allowances made to a participant (wages and commission in this question) should be recorded as expenses of the joint venture in the relevant joint venture account.

➤ Workings have been provided for the commission.

➤ The intermediate settlement between the participants has been recorded.

➤ The memorandum joint venture account includes all the income and expense items.

➤ The final settlement between the participants is for the same amount and closes the joint venture accounts.

1 State **two** reasons why a business might sell goods on consignment.

2 Explain what is meant by del credere commission.

3 A small business has been using a manual accounting system, but the owner has been advised by a friend to switch to using a computerised system. Discuss the advice given by the owner's friend.

4 H Limited sells goods on consignment through its agent, Amal, in Colombo, Sri Lanka. Amal is entitled to a 3 per cent commission and a 2 per cent del credere commission on all sales. The following transactions occurred during the financial year ended 31 December 2016.

January	Opening balance: amount outstanding from 2015, $8 450.
February	Pro forma invoice was sent to Amal for goods cost $20 000.
	H Limited paid shipping costs $870 and insurance $480.
March	H Limited received $6 000 in part settlement of the amount outstanding.
	Amal paid landing charges $850 and storage costs $300.
July	Amal sent an account sales reporting the sale of three-fifths of the consignment sent in February with a mark-up of 50 per cent.
October	Amal sent a second account sales reporting the sale of recording the sale of goods, $6 000 which included a mark-up of 50 per cent.
	Amal paid $640 to deliver goods sold to customers.
November	Amal had to write off a customer's account as an irrecoverable debt. The customer owed $1 600.
	H Limited received $12 500 from Amal in part settlement of the amount due.
December	End of year financial statements were prepared by both H Limited and Amal. Unsold consignment inventory was valued.

(a) Calculate the value of the unsold inventory at 31 December 2016.

(b) Prepare the following accounts in the books of H Limited: (i) Consignment to Amal, Colombo; (ii) Amal, Colombo.

(c) Prepare the following account in the books of Amal, Colombo: H Limited.

(d) Calculate the profit made by Amal on the consignment from H Limited during the year ended 31 December 2016.

5 Rana, Bilaal and Lee recently agreed to form a joint venture in which profits and loss would be shared in the ratio Rana:Bilaal:Lee, 2:2:1. They will be selling T-shirts which have been designed by Rana. The three participants are considering forming a formal partnership in future in their joint venture is successful. It was agreed that the joint venture would last for three months: 1 August to 31 October 2016.

The following information is available:

1 August	Lee paid for the production of 500 T-shirts featuring five different designs, $1 600.
16 August	Bilaal paid for the rent of some storage facilities for the months of August–October, $410.
18 August	Rana paid advertising costs of $220.
31 August`	Lee received $1400 from a retailer who purchased 25 per cent of the T-shirts.
15 September	It was agreed that Rana should receive $600 for her work in designing the T-shirts.
28 September	Another retailer purchased 40 per cent of T-shirts and agreed to pay Bilaal sufficient to give a gross margin of 50 per cent.
15 October	Delivery charges of $290 were paid by Rana.
20 October	It was agreed that Lee should receive a wage of $300 and Bilaal a wage of $200 for their work in delivery goods to customers.
21 October	A further 25 per cent of the T-shirts were sold to other retailers and $1800 was received by Rana.
28 October	Bilaal agreed to take over the unsold T-shirts at cost price.
31 October	The participants settled any amounts owing to each other.

Prepare the joint venture accounts in the books of (i) Rana, (ii) Bilaal, (iii) Lee, and a memorandum joint venture account.

Key topics

➤ ratio analysis and interpretation

➤ gearing

➤ investment ratios

➤ further liquidity ratios

➤ the net working assets/revenue ratio

➤ greatest challenges

➤ important techniques.

✓ What you need to know

This unit is concerned with three topics:

➤ calculating additional ratios required by potential investors in a business

➤ analysing and evaluating the results of ratios and drawing conclusions

➤ making recommendations to potential investors.

16.1 Ratio analysis and interpretation

Equity shareholders are the owners of a limited company and are therefore a particularly important group of stakeholders. This group of investors together with others who are potential investors in equity shares are concerned with:

➤ the company's liquidity

➤ the market value of the shares

➤ the degree of risk within the business.

Ratios used for analysis by investors and potential investors include:

➤ gearing

➤ investment ratios

➤ further liquidity ratios.

16.2 Gearing

The ratio for calculating gearing is:

$$\frac{\text{Fixed cost capital}}{\text{Total capital}} \times 100$$

➤ Fixed cost capital: non-current liabilities + issued preference share capital

➤ Total capital is: fixed cost capital + issued ordinary share capital + all reserves + fixed cost capital

Key term

Gearing: the relationship between fixed cost capital and total capital.

High-geared companies

A high-geared company will have a relatively high proportion of fixed-return finance.

A company could lower its gearing ratio by redeeming debentures, issuing new ordinary shares or building up reserves (for example by retaining profits).

Low-geared companies

A low-geared company will have a relatively low proportion of fixed-return finance and a relatively large proportion of finance provided by equity shareholders.

A company could increase its gearing ratio by issuing debentures, issuing new preference shares or buying back ordinary shares

Income gearing ratio

It is possible to measure the impact of interest payments on a company's profit by using the following ratio:

$$\frac{\text{Interest expense}}{\text{Profit before interest and tax}} \times 100$$

The ratio measures how easily a company has covered the cost of long-term borrowing from profits.

16.3 Investment ratios

▼ Table 16.1 Investment ratios

Ratio	Formula	What does it tell you?
Earnings per share (EPS)	$\dfrac{\text{Profit after tax and preference dividends}}{\text{Number of ordinary shares}}$	How much profit can be attributed to each issued ordinary share? (The Companies Act 1985 requires this ratio to be published in its annual financial statements.)
Dividend yield	$\dfrac{\text{Dividend paid per share}}{\text{Market price per share}} \times 100$	Gives the real rate of return on an investment in ordinary shares
Dividend cover	$\dfrac{\text{Profit available to pay ordinary dividend}}{\text{Ordinary dividend paid}}$	➤ The likelihood the dividends could be maintained if profits fall ➤ The directors approach to dividend payments (generous or mean)
Dividend per share	$\dfrac{\text{Ordinary dividend paid}}{\text{Number of issued ordinary shares}}$	The rate of dividend (to be compared with previous years)
Price/ earnings (P/E)	$\dfrac{\text{Market price per ordinary share}}{\text{Earnings per share}}$	Gives the number of years' earnings investors are prepared to pay to purchase a share

16.4 Further liquidity ratios

Working capital cycle (in days)

The ratio is calculated as follows:

Trade receivable turnover (in days) + inventory turnover (in days) – trade payable turnover (in days)

The ratio provides information about how long it takes to receive cash from selling goods taking account of how quickly goods are sold adding on the time taken for credit customers to pay less the time taken to pay suppliers for the goods.

If the ratio indicates a long delay this is likely to be a weakness for the business because it will indicate that there is likely to be cash flow problems as the business tries to cover all the other cash outflows while waiting for this major cash inflow. For example, problems could arise if loan repayments have to be delayed, if directors decide to curtail dividends, or if there are insufficient cash funds to replace inventory. If the ratio indicates a short delay, the chance of cash flows problems is reduced and this would be a strength for the business.

If a business has to try to shorten the working capital cycle it could:

➤ delay paying suppliers (but maybe risk the withdrawal of credit facilities)

➤ put pressure on credit customers to pay more quickly (but maybe risk losing customers who can obtain more favourable credit terms elsewhere)

➤ reduce inventory levels by purchasing less (but this could reduce the choice available to customers, or risk shortages of certain items and therefore not be able to meet demand).

16.5 Net working assets/revenue ratio

The ratio is calculated as follows:

$$\frac{\text{Net working assets}}{\text{Revenue}} \times 100 \text{ i.e.}$$

$$\frac{(\text{Inventories} + \text{Trade receivables} - \text{Trade payables})}{\text{Revenue}} \times 100$$

It gives an indication of the percentage of revenue which is net working assets.

16.6 What are the greatest challenges?

➤ Remembering the formula for each of the ratios

➤ Explaining the significance of each ratio

➤ Writing a report on a business's performance from the point of view of an investor

➤ Making a choice of the best investment and justifying the decision.

16.7 Review some important techniques

> **▼ Example 16.1** Writing about gearing from an investor's point of view
>
> Cheng is considering investing in ordinary shares in either C Limited or D Limited. He has been advised to check each company's gearing ratio. The following information is available.
>
> Summarised Statements of financial position at
> 31 December 2016
>
> | | C Limited | D Limited |
> |---|---|---|
> | | $m | $m |
> | Total assets | 440 | 960 |
> | Equity | | |
> | Issued share capital | 215 | 290 |
> | Share premium | 45 | 50 |
> | Retained earnings | 82 | 71 |
> | Total equity | 342 | 431 |
> | Non-current liabilities: | | |
> | 8% Debentures | 84 | 510 |
> | Current liabilities | 14 | 29 |
> | Total equity and liabilities | 440 | 960 |
>
> It is forecast that it is unlikely that these companies will be able to increase profits in the near future due to uncertain economic conditions.

Step 1: calculate gearing ratios

	C Limited	**D Limited**
Gearing ratio is $$\frac{\text{Fixed cost capital}}{\text{Total capital}} \times 100$$	$\frac{84}{426} \times 100$ $= 19.72\%$	$\frac{510}{931} \times 100$ $= 54.78\%$

> **Remember**
>
> State the formula first when asked to calculate ratios, then select the relevant data before giving the final result.

Step 2: explain ratios

C Limited is a low-geared company as the ratio is (significantly) less than 50 per cent. Its fixed cost capital (in the form of debentures) is a small proportion of the total long-term finance as there is a higher proportion of finance supplied by ordinary shareholders.

D Limited is a high-geared company as the ratio is more than 50 per cent. Its fixed cost capital (in the form of debentures) is a substantial proportion of the total long-term finance and a far lower proportion of finance is supplied by ordinary shareholders.

Step 3: explain significance of ratios to the potential investor

C Limited would be described as a low-risk company. This is because debenture interest, which must be paid each year, will be relatively low and so has less impact on profits. As a result, it is less likely that dividends will be affected if profits should decrease. Low-geared companies generally find it easier to increase borrowing, if required, since existing levels of borrowing are low.

However, D Limited would be described as a high-risk company. This is because debenture interest, which must be paid each year, will relatively high and so will have a more significant impact on profits. If profits should fall significantly it is likely that dividends could be sharply reduced, and in severe cases the company could be forced into liquidation. Furthermore, high-geared companies generally are far less likely to be able to increase borrowing as existing levels of borrowing are high, and banks or other lenders are likely to question why ordinary shareholders are not being asked to invest more.

Step 4: make a recommendation (if required by the question)

As profits are unlikely to increase, the recommendation is that Cheng invests in C Limited as it is low-risk.

▼ **Example 16.2** Calculating investment ratios

The following information is available for Q Limited:

For the year ended 31 December 2016

	$000	At 31 December 2016	
Profit from operations	540		$000
Finance charges	(36)	Total assets	4232
Profit before taxation	504	Equity	
Taxation	(64)	Ordinary shares of $1 each	2400
Profit for the year	440	Share premium	300
		Retained earnings	850
		Total equity	3550
		Non-current liabilities	
		6% Debentures	600
		Current liabilities	82
		Total equity and liabilities	4232

Additional information:

The directors of Q Limited propose a dividend of $0.15 per share for the year ended 31 December 2016.

At 31 December 2016 the market price of an ordinary share in Q Limited is $2.80.

Ratio	Formula	Calculation
Income gearing	$\dfrac{\text{Interest expense}}{\text{Profit before interest and tax}} \times 100$	$\dfrac{36}{540} \times 100 = 6.67\%$
Earnings per share (EPS)	$\dfrac{\text{Profit after tax and preference dividends}}{\text{Number of ordinary shares}}$	$\dfrac{440}{2400} = \$0.18$
Dividend yield	$\dfrac{\text{Dividend paid per share}}{\text{Market price per share}} \times 100$	$\dfrac{\$0.15}{\$2.80} \times 100 = 5.36\%$

Dividend cover	Profit available to pay ordinary dividend / Ordinary dividend paid	$\frac{440}{360} = 1.22$ times
Price/earnings (P/E)	Market price per ordinary share / Earnings per share	$\frac{\$2.80}{\$0.18} = 15.56$

Remember

Check how many decimal places are required for the answer.

✗ Common error

Don't forget to label each result with the correct descriptor.

Revision checklist

I can:

➤ recall the formula for gearing, investment ratios and additional liquidity ratios ☐

➤ describe what each of the ratio tells an investor about performance ☐

➤ compare the performance of business from an investor's viewpoint ☐

➤ make recommendations about which investment is most suitable. ☐

 Raise your grade

The directors of W Limited wish to invest in either E Limited or F Limited. The following information is available:

	E Limited	F Limited
Gearing	61%	22%
Earnings per share	$0.41	$0.63
Dividend cover	4.23 times	6.71 times
Dividend yield	10%	5.71%
Price earnings ratio	6.10	5.56

Advise the directors whether they should invest in E Limited or F Limited.

Student answer

❶

E Limited's gearing ratio is high compared to F Limited's ratio which is low. ❷

F Limited has the higher earnings per share, in fact it is 50 per cent higher. ❷

F Limited's dividend cover is higher than E Limited. ❷

E Limited's yield is higher which means the return on the investment is higher.

The price earnings ratio of both companies is similar. ❸

I think W Limited should invest in F Limited. ❹ ❺

How to improve this answer

① Statements are very brief and could be more effectively organised to make the comparisons clearer

② Comparisons are correct but the significance of the comparisons is not explained

③ Not clear which company is performing better

④ Decision is made but no justification is provided

⑤ No mention is made of wider factors which should be considered

Model answer

E Limited performs better in regard to:

- yield which indicates that the return on the investment is higher

- price/earnings ratio (though it is only very marginally higher than that of F Limited) which indicates that investors are prepared to pay a little more for shares in E Limited.

F Limited performs better in regard to:

- earnings per share (eps is 50 per cent higher than for E Limited) which indicates that this is the safer investment

- dividend cover (again about 50 per cent greater than for E Limited) which indicates that it more likely that the current dividend policy could be sustained if profits should fall (but could also indicate that F Limited is less inclined to proposed dividend payments than E Limited).

E Limited is high-geared and F Limited is low-geared. This means that E Limited is a high risk company and F Limited is low risk. F Limited is, therefore, more vulnerable to a fall in dividend payments should profits fall, whereas a fall in profits in E Limited would have a smaller effect on dividend payments. However, if profits in E Limited should increase, investors in ordinary shareholders could see a significant increase in dividends as there are relatively few of them to share the profits; if profits increase in F Limited this is not likely to have much impact on dividend payments as there are relatively many more shareholders to share in the increased profits.

The directors need also to consider:

- ratio trends over recent years

- forecasts about growth and economic climate

- whether they are prepared to take risks with the investment.

The picture which emerges is somewhat mixed. However, F Limited is recommended as it is a safer investment where current dividend levels are more likely to be maintained in the future.

Note: other creditworthy points could be made and would be rewarded.

★ **Exam tips**

Key features of the model answer

➤ The information is presented more clearly.

➤ Each ratio is compared and the significance to the ratio explained.

➤ Clarification is given as to which company is performing better for each ratio.

➤ Full explanations are given (for example of the implication of the gearing ratio).

➤ Other factors for consideration are mentioned.

➤ A recommendation is made and a justification is provided.

Exam-style questions

1 State which investment ratio must be published by a limited company in its annual financial statements.

2 Describe what is meant by the term 'low-geared company'.

3 Explain the usefulness of the dividend cover ratio to a potential investor.

4 Farzana has recently inherited $100 000 which she wishes to invest in shares in a limited company. She has decided to choose between buying shares in U limited or V Limited.

The following information is available concerning U Limited:

For the year ended 31 December 2016

	$m
Profit from operations	4.25
Finance charges	(0.24)
Profit before taxation	4.01
Taxation	(0.89)
Profit for the year	3.12

At 31 December 2016

	$m
Total assets	10.83
Equity	
Ordinary shares of $1 each	4.00
Share premium	1.25
Retained earnings	1.37
Total equity	6.62
Non-current liabilities	
8% Debentures	3.00
Current liabilities	1.21
Total equity and liabilities	10.83

The following ratios have been calculated for V Limited:

Dividend yield	5.66%
Dividend cover	7.42 times
Earnings per share	$0.23
Price earnings ratio	21.52
Gearing	62%
Income gearing	18%

Additional information

	U Limited	V Limited
Current market price of $1 ordinary share	$3.82	$4.95
Proposed dividend per share at 31 December 2016 (no interim dividends per paid during the year ended on this date)	22p	28p

Required

(a) Calculate the following ratios (to two decimal places) for U Limited from the income statement and statement of financial position:

(i) Dividend yield

(ii) Dividend cover

(iii) Earnings per share

(iv) Price earnings ratio

(v) Gearing

(vi) Income gearing

(b) Advise Farzana whether she should invest in shares in U Limited or V Limited. Give reasons for your decision.

Key topics

➤ the concept of activity-based costing

➤ advantages and disadvantages of activity-based costing (ABC)

➤ comparison of activity-based costing with absorption costing

➤ greatest challenges

➤ important techniques.

✓ What you need to know

This unit reviews apportioning overheads using activity-based costing techniques, calculating the total cost of a unit, calculating the value of inventory, the advantages and disadvantages of activity-based costing and demonstrating the effect of different methods of overhead absorption on profit.

17.1 The concept of activity-based costing

This technique focuses on discovering the activities which cause overheads to occur and using this information to allocate costs.

Overheads are allocated to different products based in relation to the extent to which each product causes activities to arise. The more a particular product causes activities to arise the greater its share of overheads.

Cost pools

Activities are first organised into a group of overheads which relate to a specific aspect referred to as cost pools which are often the same as the organisation's departments or service centres. An example of a cost pool is machine set ups.

> **Key term**
>
> Cost pool: the location of a group of costs.

Each cost pool should gather together all related overheads which vary according to the same activity. Example, delivery to customers:

➤ related costs are: depreciation of vehicles, fuel costs, repairs and maintenance costs;

➤ they vary with the same activity: the mileage involved in each delivery.

Cost drivers

Cost drivers are activities which cause costs to arise. Example: the number of times each machine has to be set-up to make particular products.

> **Key term**
>
> Cost driver: activities that cause costs to arise.

Further examples:

Cost pool	Cost driver
Issuing materials from stores	Number of materials requisitions
Transfers of partly finished goods between machines and between production departments	Number of transfers
Quality checks	Number of quality inspections
Machine maintenance	Number of maintenance hours
Packaging	Number of units
Delivery to customers	Number of deliveries

Valuation of inventories

Unsold units are valued on the basis of total cost per unit, i.e. direct costs per unit plus overheads per unit.

17.2 Advantages and disadvantages of activity-based costing

Advantages

➤ More accurate allocation of costs avoiding the arbitrariness of more traditional techniques. Traditional techniques are less likely to be relevant in age of more sophisticated technology.

➤ Provides a better basis for decision making, for example pricing and assessing the viability of certain products because information is more reliable.

➤ Is equally useful in service providers as well as manufacturing organisations.

Disadvantages

➤ Expensive to introduce the technique because of the very detailed work needed in reviewing every aspect of the production process.

➤ Additional expense incurred because specialists usually required to set up the system, training costs for those who operate the system, additional costs of keeping the system up to date.

17.3 Comparison of activity-based costing with absorption costing

Activity-based costing is more reliable and precise than absorption costing because it takes a detailed account of the activities which cause overheads to arise; whereas absorption costing takes a rather broad approach of apportioning overheads based on machine hours or labour hours which may not be completely relevant in some production departments. As a result, activity-based costing is far less likely to lead to the over-absorption or the under-absorption of overheads and so forecast profits are less subject to alteration.

17.4 What are the greatest challenges?

➤ Understanding the basic concepts on which activity-based costing is based.

➤ Making accurate calculations of the overhead allocation rates.

➤ Using overhead allocation rates when calculating the total cost of a product.

17.5 Review some important techniques

Allocating overheads

▼ **Example 17.1** Using information about cost pools and cost drivers to determine the allocation of overheads

A company produces three products: X, Y and Z. The following information is available concerning cost pools and cost drivers.

		Overheads	Product X	Product Y	Product Z
Cost pool	Cost driver		per month		
Machine set ups	Number of sets up required for each product	$34 500	1 set-up	2 set-ups	3 set-ups
Quality checks	Number of quality check for each product	$18 000	1 check	1 check	2 checks
Total overheads		$52 500			

> **Exam tip**
>
> When you have allocated overheads it is recommended that you do a quick check to make sure the allocations add up to the correct total. In the illustration $5400 + $12 600 + $34 500 do add up to the correct total of $52 500.

During December 2016 production was: X 2000 units; Y 3000 units: Z 5000 units.

Step 1: calculate an overhead allocation rate

	Product X 2 000 units	Product Y 3 000 units	Product Z 5 000 units	Total	Total overheads	Overhead allocation rate
Machine set ups	2000	6 000	15 000	23 000	$34 500	$1.50 per set-up
Quality checks	2000	3 000	10 000	15 000	$18 000	$1.20 per check

Step 2: allocate overheads to each product

	Product X		Product Y		Product Z	
Machine set-ups	2 000 × $1.50	$3 000	6 000 × $1.50	$9 000	15 000 × $1.50	£22 500
Quality checks	2 000 × $1.20	$2 400	3 000 × $1.20	$3 600	10 000 × $1.20	$12 000
TOTALS		$5 400		$12 600		$34 500

> **Remember**
>
> To set out detailed workings to show how you arrived at the overheads allocated to each product.

▼ **Example 17.2** Calculating the total cost of one unit

To calculate the cost of one unit requires the addition of direct costs per unit plus overheads costs established using the technique in Illustration 1.

Using the data from Example 17.1 and included figures for direct materials and direct labour (which are as shown in the table below), the total cost of one unit for each product can be calculated as follows.

Step 1: adding total direct costs and total overheads

	Product X	Product Y	Product Z
	$	$	$
Direct materials	7 300	14 800	26 200
Direct labour	6 400	12 100	22 300
Total overheads	5 400	12 600	34 500
Total cost	19 100	39 500	83 000

Step 2: Dividing total cost by the number of units produced for each unit:

	Product X	Product Y	Product Z
	$	$	$
Total cost	$19 100	$39 500	$83 000
Number of units	2 000	3 000	5 000
Cost per unit	$9.55	$13.17	$16.60

Revision checklist

I can:

➤ define key terms associated with activity-based costing ☐

➤ discuss the benefits and drawbacks of changing from a traditional costing method to activity-based costing ☐

➤ calculate overhead allocation rates and the total cost of a product ☐

➤ give advice and make recommendations based on evidence from the application of activity-based costing. ☐

↑ Raise your grade

G Limited manufacture two products: quams and strams. The following information is available for production during November 2016.

	Direct materials	Direct labour	Production (units)
	$ per unit	$ per unit	units
Quams	$8.20	$7.40	6 000
Strams	$11.90	$10.30	4 000

Cost pool	Cost driver	Overhead cost per month	Quams	Strams
Issues from stores	Number of material requisitions	$16 500	5 per unit	9 per unit
Transfers of partly finished goods between departments	Number of transfers	$17 600	4 per unit	5 per unit
Quality checks	Number of checks	$16 800	2 per unit	4 per unit

The directors wish to achieve a gross profit margin of 20 per cent on each unit.

The directors have considered increasing the selling price of strams.

(a) Calculate the selling price of each unit.

(b) Discuss the idea of increasing the selling price of strams.

Student answer

Overhead allocation rates ❶

	Quams	Strams	Allocation rate
Issues from stores	30 000	36 000	$0.25 ❷
Transfers between departments	24 000	20 000	$0.40 ❷
Quality checks	12 000	16 000	$0.60 ❷

Allocation of overheads ❶

	Quams	Strams
	$	$
Issues from stores	7500	9 000
Transfers between departments	9 600	8 000
Quality checks	7200	9 600
Total overheads ❸	24 300	26 600

Selling price per unit

Quams: $24 300 ❹ × 120% ❺ = 29 160/4000 = $7.29

Strams: $26 600 ❹ × 120% ❺ = 31 920/6000 = $5.32

The directors could consider increasing the price of strams as this will increase profits on this product. ❻

How to improve this answer

❶ There is a lack of detailed workings throughout the answer. This could have been a high-risk strategy because any minor errors in calculations could have resulted in a considerable loss of marks because there would have been no evidence to show. whether there was a lack of understanding or the technique or just a minor slip-up. (As it happens the candidate did calculate the allocation of overheads correctly.)

❷ The allocation rates are correct but lack a descriptor: $0.25 per issue, $0.40 per transfer, $06.0 per check.

❸ There is no evidence that the candidate took the precaution to check that the total overheads for each product when totalled equalled the overall total for overheads ($50 900).

❹ The direct costs have been ignored.

❺ The candidate has applied a mark-up of 20 per cent, whereas the question required a gross margin of 20 per cent.

❻ The candidate has provided a very limited response in part (b) and has ignored the fact that a discussion was required detailing the benefits and drawbacks of the possible increase in price.

Model answer

Calculation of overhead allocation rates:

	Quams	Strams	Totals	Total overheads	Allocation rate
Issues from stores	6 000 × 5 = 30 000	4 000 × 9 = 36 000	66 000	$16 500	$0.25 per issue
Transfers between departments	6 000 × 4 = 24 000	4 000 × 5 = 20 000	44 000	$17 600	$0.40 per transfer
Quality checks	6 000 × 2 = 12 000	4 000 × 4 = 16 000	28 000	$16 800	$0.60 per check
Total overheads				$50 900	

Allocation of overheads:

	Quams		Strams	
Issues from stores	30 000 × $0.25	$7 500	36 000 × $0.25	$9 000
Transfers between departments	24 000 × $0.40	$9 600	20 000 × $0.40	$8 000
Quality checks	12 000 × $0.60	$7 200	16 000 × $0.60	$9 600
Total overheads		$24 300		$26 600
Overall total overheads:	$24 300 + $26 600 = $50 900			

Calculation of total cost and sales

	Quams	Strams
	$	$
Materials		
Quams: 6 000 x $8.20	49 200	
Strams: 4 000 x $11.90		47 600
Labour		
Quams: 6 000 x $7.40	44 400	
Strams: 4 000 x $10.30		41 200
Overheads	24 300	26 600
Total cost	117 900	115 400
Profit (20% on selling price; 25% x cost price)	29 475	28 850
Total sales	147 375	144 250

Selling price per unit

	Quams	Strams
	$	$
Total sales	147375	144250
Number of units	6000	4000
Selling price per unit	$24.56	$36.06

If the directors increase the price of strams this could lead to increased profits. However, the directors need to consider if there are any competitors selling strams or a virtually identical product and the competitor's pricing policy. If there is a competitor and the price of the equivalent product is much higher there could be an opportunity to increase prices and make more profits. However, if the competitor's prices are similar any increase in prices could lead to a fall in demand and a consequent loss of profits. If there is no competition it may be possible to increase prices without affecting demand, but this will be dependent on how essential the product is to potential customers.

⭐ **Exam tips**

Key features of the model answer

➤ Detailed workings have been provided at each stage of the process to ensure that anyone checking the answer can see exactly how each element was achieved.

➤ Each allocation rate has been given a descriptor.

➤ There is a built-in check that the total overheads allocated to each product equal the overall total for overheads.

➤ The calculation of the total cost of each product includes direct costs.

➤ A gross profit margin of 20 per cent (equating to a 25 per cent mark-up) has been correctly applied.

➤ There is a detailed discussion of the proposal to increase prices, pointing out the possible benefits and the circumstances in which the proposal could work, but also setting out potential drawbacks and the circumstances in which this proposal could be counter-productive.

1 State **two** reasons why a business might decide to adopt activity-based costing.

2 Describe the key features of a cost pool.

3 Explain the term cost driver.

4 G Limited manufactures two products: beros and keros. The following information is available:

Direct costs per unit:

	Beros	Keros
	$	$
Direct materials	17.00	23.00
Direct labour	9.00	12.00

Annual production and sales are:

Beros	14 000 units
Keros	9 000 units

Total overheads are $260 050 per annum.

The company has the following cost pools and cost drivers.

Cost pool	Cost driver	Overhead cost per annum	Details about each product	
			Beros	Keros
Issues from stores	Number of times materials are taken from stores to the production lines.	$89 250	4 issues	7 issues
Machine set-ups	Number of times machines require to be set up for each stage in the manufacture of products.	$94 000	11 times	9 times
Quality checks	Number of inspection checks during the production process.	$76 800	3 times	6 times

(a) Calculate the overhead allocation rate for each product.

(b) Calculate the total cost per unit of each product.

Additional information:

Selling prices per unit are currently:

Beros	$42.50
Keros	$65.50

(c) Calculate the annual profit to be made on each product.

The market for beros is very competitive and the directors have become aware that a rival company is likely to cut prices of this product in the near future. The directors are considering switching production from beros to keros.

(d) Discuss the proposed change in production from beros to keros.

Key topics

- ➤ what is budgetary control?
- ➤ types of functional budget
- ➤ the master budget
- ➤ flexing the budget
- ➤ greatest challenges
- ➤ important techniques.

✓ What you need to know

This unit reviews budgets including preparing a range of functional budgets, preparing a master budget and flexing budget statements.

18.1 What is budgetary control?

Budgetary control requires the preparation of functional budgets each of which is designed to control resources so that the business's objectives can be achieved.

18.2 Types of functional budget

There are a number of types of functional budgets:

Sales budget

Sets out for each budget period the expected number of units to be sold and the expected sales value.

Production budget

Identifies the required level of production in units for each budget period taking account of expected inventory levels and expected sales.

Labour budget

Sets out the required labour hours and labour costs for each budget period based on expected production levels; surplus hours or any shortfall in hours are also identified.

Purchases budget

Sets out the cost of materials required each budget period based on budgeted production

Trade receivables budget

Identifies the amount outstanding from trade receivables at the end of each budget period based upon the amount outstanding at the beginning of each budget period, the sales for each budget period and the receipts for each budget period.

Trade payables budget

Identifies the amount still owing to trade payables at the end of each budget period based upon the amount owing at the beginning of each budget period, the purchases for each budget period and the payments for each budget period.

Key terms

Budget: a short-term financial plan.

Functional budget: a short term financial plan for a particular business operation.

✗ Common error

Don't forget that a production budget is in units and don't attempt to give a monetary value to each budget period's production.

Cash budget

Identifies the cash balance at the end of each budget period and is based upon the cash inflows for each budget period less the cash outflows for that period. It provides information in good time about cash surpluses which could be put to better use and cash shortages which need to be covered by making an overdraft arrangement or by applying for a bank loan.

18.3 The master budget

Details from each functional budget are used to prepare a master budget which is a forecast income statement and a forecast statement of financial position.

Budgets are often used to evaluate the performance of managers comparing actual performance against targets for each manager's area of responsibility. Where there is a difference – called a variance – an effort is made to discover the underlying causes, so that beneficial results (favourable variances) can be maintained, or the reasons for unfavourable results (adverse variances) discovered and wherever possible remedied.

Limiting factors

When preparing budgets managers will find that they are constrained by a shortage of some kind. This limiting factor usually becomes the starting point for preparing the functional budgets. Typical limiting factors include:

➤ Demand for products
➤ Raw material availability
➤ Space available for production
➤ Storage capacity
➤ Labour hours available
➤ Cash or finance available.
➤ Machine hours available

18.4 Flexing the budget

A business could prepare just one set of budgets based on certain objectives and plans. However, this would be of very limited use and could be potentially misleading if there is a range of possible outcomes depending on actual levels of activity. It is usual therefore to provide more than one set of budgets to cover likely outcomes and this is known as flexing the budget. Flexed budgets are valuable because they:

➤ ensure managers compare actual results with the relevant budget information, comparing like with like

➤ avoid decisions being based on misleading information.

18.5 What are the greatest challenges?

➤ Remembering the purpose of each functional budget

➤ Preparing each functional budget using the correct layout

➤ Taking account of every detail provided in preparing budgets

➤ Giving advice and making recommendation based on budget outcomes.

18.6 Review some important techniques

Preparing a production budget

▼ **Example 18.1**

B Limited's sales budget for five productions periods in the near future is as follows:

	Period 1	Period 2	Period 3	Period 4	Period 5
Sales (units)	6300	6700	7100	6600	6500

Inventory at the beginning of period 1 is forecast to be 1260 units. The company's policy is to maintain the closing inventory (in units) at 20 per cent of the next period's sales. However, storage capacity is a limiting factor. The maximum number of units that can be stored during the budget period is 1350 units.

> **Remember**
>
> To find the figure for production add together sales and closing inventory and then deduct opening inventory.

The production budget for periods 1–4 inclusive is as follows:

Production budget for periods 1–4 (units)				
	Period 1	Period 2	Period 3	Period 4
Sales	6300	6700	7100	6600
Opening inventory	(1260)	(1340)	(1350)	(1320)
Closing inventory	1340	1350*	1320	1300
Production	6380	6710	7070	6580

*closing inventory should have been 1420 units in line with the company's policy, but the limiting factor of storage capacity has been applied (1350 units).

Preparing a labour budget

▼ **Example 18.2**

B Limited (see Example 18.1) has 33500 labour hours available each period in normal working hours. The labour rate of the budgetary period will be $9 per hour in normal working hours; in overtime conditions direct labour is paid 'time and a half'. Each unit of production takes 5 labour hours.

Taking account of the production budget (Illustration 1) the labour budget for periods 1–4 inclusive is as follows:

Labour budget for periods 1–4 inclusive				
	Period 1	Period 2	Period 3	Period 4
Production	6380 units	6710 units	7070 units	6580 units
Labour required	31900 hrs	33550 hrs	35350 hrs	32900 hrs
Normal working hours available	33500 hrs	33500 hrs	33500 hrs	33500 hrs
Labour cost (normal working hours) (× $9 per hour)	$287100	$301500	$301500	$296100
Surplus hours	1600 hrs			600 hrs
Shortfall (overtime) hours		50 hrs	1850 hrs	
Overtime cost (× $9 × 1.50 per hour)		$675	$24975	
Total labour cost	$287100	£302175	$326475	$296100

Managers will need to consider how best to use the surplus hours in Period 1 and Period 4.

Preparing a trade receivables budget

▼ Example 18.3

C Limited's sales budget for periods 1–4 is as follows:

	Period 1	Period 2	Period 3	Period 4	Period 5
Sales (units)	500	500	600	600	500

> **Remember**
>
> All budgets should be given a full heading.

Each unit is sold for $10. It is expected 20 per cent of each period's sales will be for cash. Seventy-five per cent of credit customers will pay in the period following sale and 25 per cent of credit customers will pay in the second period after the sale.

A trade receivables budget is required for each of the periods 3–5 inclusive.

The budget is as follows:

Trade receivables budget for each of period 3–5 inclusive			
	Period 3	Period 4	Period 5
	$	$	$
Opening balance	5 000	5 800	6 000
Credit sales (W1)	4 800	4 800	4 000
Receipts (W2):			
From previous period	3 000	3 600	3 600
From two periods before	1 000	1 000	1 200
Closing balance (W2)	5 800	6 000	5 200

W1: calculation of credit sales

	Period 1	Period 2	Period 3	Period 4	Period 5
Sales (units)	500	500	600	600	500
Selling price per unit	10	10	10	10	10
Total sales	5 000	5 000	6 000	6 000	5 000
Credit sales (80% of total sales)	4 000	4 000	4 800	4 800	4 000

W2: calculations of receipts each period, unpaid amounts, closing balances

Period	Sales $	cash after 1 period (75%) $	cash after 2 periods (25%) $	total receipts $	unpaid this period's sales $	unpaid from last period's sales $	closing balance $
1	4 000				4 000	1 000	
2	4 000	3 000		3 000	4 000	1 000	5 000
3	4 800	3 000	1 000	4 000	4 800	1 000	5 800
4	4 800	3 600	1 000	4 600	4 800	1 200	6 000
5	4 000	3 600	1 200	4 800	4 000	1 200	5 200

(The shaded section shows data required for the budget periods 3, 4 and 5.)

Preparing a trade receivables (and trade payables) budget can be more complicated than it at first appears. Detailed workings can help considerably to sort out what happens and when.

Preparing a master budget

▼ **Example 18.4**

The financial statements of G Limited's for the year ended 31 December 2016 were as follows.

Summarised income statement
for the year ended 31 December 2016

	$000
Revenue	1 000
Cost of sales	(500)
Gross profit	500
Distribution costs	(127)
Administration expenses	(65)
Profit from operations	308
Finance costs	(8)
Profit before taxation	300
Taxation	(60)
Profit for the year	240

Statement of changes in equity (extract)
for the year ended 31 December 2016

	Retained earnings $000
Balance, 1 January 2016	104
Profit for the year	240
Dividends paid	(96)
Balance, 31 December 2016	248

Statement of financial position at 31 December 2016

Assets	$000
Non-current assets	
Property, plant and equipment	882
Current assets	
Inventory	31
Trade receivables	72
Cash and cash equivalents	11
	114
Total assets	996
Equity and liabilities	
Equity	
Ordinary shares of 50c each	480
Share premium	60
Retained earnings	248
	788
Non-current liabilities	
8% Debentures (2025)	100

Current liabilities		
Trade payables		48
Taxation		60
		108
Total equity and liabilities		996

The following budget plans have been produced for the year ending 31 December 2017.

1. Revenue is expected to increase by 20% by reducing the gross profit margin by 5%.

2. Distribution costs are expected to decrease by $11 000 and administration expenses are expected to increase by $21 000.

3. There will be no change in finance costs and taxation will remain at the same percentage of profit before tax.

4. A rights issue is planned for 1 April 2016 consisting of 2 shares for every 3 shares in issue at that date at a price of 70c per share. The directors are confident that the issue will be fully subscribed.

5. The funds raised by the share issue will be used to purchase additional plant and equipment.

6. The annual depreciation charge is expected to be $96 000 and will be divided between distribution costs and administration expenses in the ratio 5:3.

7. Dividends of 10p per share are expected to be paid on 1 September 2017 on all shares in issue at that date.

8. It is expected that the inventory will be increased to $42 000 by 31 December 2017

9. All purchases and sales will be on credit. The trade payables payment period is expected to be 32 days and trade receivables collection period is expected to be 30 days (calculations using these ratios should be made to the nearest $000.)

Step 1: preparing the budgeted income statement:

Budgeted Income Statement
for the year ending 31 December 2017   Workings

	$000	
Revenue	1 200	1000 × 120% = 1200
Cost of sales	(660)	
Gross profit	540	1200 × (50% − 5%) = 540
Distribution costs	(116)	127 − 11 = 116
Administration expenses	(86)	65 + 21 = 86
Profit from operations	338	
Finance costs	(8)	
Profit before taxation	330	
Taxation	(66)	330 × 20% = 66
Profit for the year	264	

Step 2: Preparing workings for the budgeted statement of financial position

W1 Rights issue: 960 000 shares × 2/3 = 640 000 shares

Issued capital increases by 640 000 × 50c = 320 000

Share premium increased by 640 000 × 20c = 128 000

Total cash raised is $448 000 (used to increase

non-current assets)

W2 Non-current assets: additions $448 000 (W1)

less depreciation charge $96 000 = increase

of $352 000

W3 Retained earnings: $248 000 – dividend paid

(10p × 1600 000 shares = $160 000) + profit

for year $264 000 = $352 000

W4 Credit purchases

Opening inventory $31 000 + Purchases –

Closing inventory $42 000 = cost of sales $660 000

hence purchases is $671 000

W5 Closing trade payables

Credit purchases (W4) $671 000 × 32/365

i.e. $58 827 rounded to $59 000

W6 Payments to trade payables

Opening balance $48 000 + purchases (W4)

$671 000 – closing trade payables (W5) $59 000

i.e. $660 000

W7 Closing trade receivables

Credit sales $1 200 000 × 30/365 = $98 630

rounded to $99 000

W8 Receipts from trade receivables

Opening balance $72 000 + Sales $1 200 000

less closing trade receivables (W7) $99 000,

i.e. $1 173 000

W9 Expense payments

Distribution $116 000 less depreciation

(5/8 × $96 000, i.e. 60 000) = $56 000

Administration $86 000 less depreciation

(3/8 × $96 000, i.e. 36 000) = $50 000

> **✗ Common error**
>
> Be careful not to overlook some detail given in the question resulting in inaccuracies in the financial statements. For example, if the change in the gross profit margin was ignored, the gross profit and profit for the year would be overstated by $60 000.

Step 3: to provide a detailed calculation of the closing balance of cash and cash equivalents

W9 Budgeted balance of cash and cash equivalents at 31 December 2017

	$000		$000
Opening balance	11	Trade payables (**W6**)	660
Trade receivables (**W9**)	1 173	Distribution costs (**W10**)	56
Rights issue (**W1**)	448	Administration expenses (**W10**)	50
		Dividends (**W3**)	160
		Taxation	60
		Additional non-current assets (**W1**)	448
		Finance charges	8
		Closing balance	190
	1 632		1 632

Step 4: to prepare the budgeted statement of financial position

Budgeted Statement of Financial Position
at 31 December 2017

Assets	$000
Non-current assets	
Property, plant and equipment (**W2**)	1 234
Current assets	
Inventory	42
Trade receivables (**W7**)	99
Cash and cash equivalents (**W9**)	190
	331
Total assets	1 565
Equity and liabilities	
Equity	
Ordinary shares of 50c each (**W1**)	800
Share premium (**W1**)	188
Retained earnings (**W3**)	352
	1 340
Non-current liabilities	
8% Debentures (2025)	100

Current liabilities	
Trade payables (**W6**)	59
Taxation	66
	125
Total equity and liabilities	1 565

Flexible budgets

▼ **Example 18.5**

P Limited budgeted income statement and actual income statement for November 2016 are as shown:

Budget income statement for November 2016		Actual income statement for November 2016	
	$000		$000
Revenue (20 000 units)	240	Revenue (22 000 units)	260
Cost of sales	(180)	Cost of sales	(199)
Gross profit	60	Gross profit	61
Overheads	(35)	Overheads	(33)
Profit for month	25	Profit for month	28

Overheads consist of variable costs of $30 000 and a fixed cost of $5 000

A superficial comparison would provide misleading information with the sales, gross profit and profit for the year all looking better in the actual results compared to the budget.

However, the budget should be flexed. This means scaling all the variable elements in the budget upwards by 10 per cent because the business actually sold 10 per cent more goods.

The comparison is now made on a like-for-like basis.

Budget income statement for November 2016		Flexed budgeted Income Statement for November 2016		Variances	
				Favourable	Adverse
			$000	$000	$000
	$000				
Revenue (20 000 units)	240	Revenue (22 000 units)	264		4
Cost of sales	180	Cost of sales	198		1
Gross profit	60	Gross profit	66		5
Overheads	35	Overheads	38	5	
Profit for month	25	Profit for month	28	0	0

The adverse variances for sales, cost of sales and gross profit now become apparent and should be subject of investigation to see if avoiding action can be taken. The favourable variance for overheads should also be investigated to see if this positive outcome can be maintained.

I can:

➤ explain the purpose and remember the correct format for each functional budget ☐

➤ prepare functional budgets ☐

➤ prepare a master budget ☐

➤ understand the reasons for, and prepare, a flexible budget. ☐

 Raise your grade

O Limited has provided the following information concerning its budget plans for the period May–August 2017.

Sales budget

April	May	June	July	August
$51 000	$54 000	$60 000	$58 500	$52 500

Fifty per cent of each month's sales are for cash. Credit customers pay one month after the month of sale and are allowed a discount of 2 per cent.

Purchases budget

March	April	May	June	July	August
$34 000	$36 000	$40 000	$42 000	$44 000	$44 000

All purchases are made on credit. 40 per cent of purchases are paid one month after purchase; the remainder are paid two months after purchase.

Additional information

• Administration and distribution costs are expected to total $10 000 in April, May and June, and then increase by 10 per cent on this figure in July, and by a further 5 per cent on the July figure in August. These expenses are paid for 80 per cent in the month they are incurred, and 20 per cent one month later. Distribution expenses included depreciation charges which are expected to be:

April	May	June	July	August
$1 000	$1 000	$1 200	$1 200	$1 400

• The directors plan to purchase additional non-current assets in June 2017 at a cost of $75 000. They plan to totally finance this capital expenditure by issuing $1 ordinary shares at a premium of 50c per share in May 2017.

• The directors expect to pay a dividend of 10 cents per share in August 2016 on all shares in issue at this date. (On 1 May 2016 the company's issued capital consisted of 600 000 shares of $1 each.)

On 1 May 2017 the balance of cash and cash equivalents is expected to be $37 000.

Prepare a cash budget for each of the months May–August 2017.

Student answer

Cash budget ❶

	May $	June $	July $	August $
Receipts				
Cash sales ❷	27 000	30 000	29 250	26 250
Trade receivables ❷ ❸	25 500	27 000	30 000	29 250
Share issue ❹		75 000		
Total receipts	52 500	132 000	59 250	55 500
Payments				
Trade payables (W1) ❺	37 600	40 800	42 800	44 000
Administration and distribution (W2) ❻	11 000	11 160	12 000	12 800
Non-current assets		75 000		
Dividends paid (W3) ❼				67 500
Total payments	48 600	126 960	54 800	124 300
Opening balance	37 000	40 900	45 940	50 390
Net cash receipts/(payments)	3 900	5 040	4 450	(68 800)
Closing balance ❽	40 900	45 940	50 390	(18 410)

		March $	April $	May $	June $	July $	August $
	Purchases	34 000	36 000	40 000	42 000	44 000	44 000
	Paid 1 month (40%)			16 000	16 800	17 600	17 600
	Paid 2 months (60%)			21 600	24 000	25 200	26 400
W1	Total payments			37 600	40 800	42 800	44 000
	Administration and Distribution						
	expense		10 000	10 000	10 000	11 000	11 550
	Depreciation		1 000	1 000	1 200	1 200	1 400
	Total expense		11 000	11 000	11 200	12 200	12 950
	Paid same month (80%)			8 800	8 960	9 760	10 360
	Paid 1 month (20%)			2 200	2 200	2 240	2 440
W2	Total payment			11 000	11 160	12 000	12 800

W3　Dividends paid

Original share capital ($600 000 + share issue $75 000) x 10 cents = 67 500

How to improve this answer

❶ The title is inadequate.

❷ There are no workings to support the cash sales and receipts from trade receivables figures. No marks are lost for the cash sales figures because they are correct. However, marks will be lost for the trade receivables receipts figures as they are incorrect.

❸ The candidate has ignored the information about discounts allowed and so the receipts from trade receivables are overstated.

❹ The share issue is recorded in the wrong month.

❺ The instructions about payments to trade payables has been misunderstood. The candidate has provided figures based on some immediate payment in the month of purchase and the remainder paid on month later.

❻ Depreciation has been added to the expense figures, whereas it should have been deducted as a non-cash item.

❼ The dividend payment has been based on the total value of the new share issue rather than just the face value.

❽ The balances and net cash receipts/payments figures are incorrect because of previous errors.

Model answer

O Limited
Cash budget for each of the four periods May–August 2017

	May	June	July	August
	$	$	$	$
Receipts				
Cash sales (**W1**)	27 000	30 000	29 250	26 250
Trade receivables (**W2**)	24 990	26 460	29 400	28 665
Share issue	75 000			
Total receipts	1 26 990	56 460	58 650	54 915
Payments				
Trade payables (**W3**)	34 800	37 600	40 800	42 800
Administration and distribution (**W4**)	9 000	8 840	9 600	10 080
Non-current assets		75 000		
Dividends paid (**W5**)				65 000
Total payments	43 800	1 21 440	50 400	1 17 880
Opening balance	37 000	1 20 190	55 210	63 460
Net cash receipts/(payments)	83 190	(64 980)	8 250	(62 965)
Closing balance	1 20 190	55 210	63 460	495

		March	April	May	June	July	August
		$	$	$	$	$	$
	Total sales		51 000	54 000	60 000	58 500	52 500
W1	Cash sales (50%)		25 500	27 000	30 000	29 250	26 250
	Credit sales (50%)		25 500	27 000	30 000	29 250	26 250
	Amount settled			25 500	27 000	30 000	29 250
W2	Receipt (98%)			24 990	26 460	29 400	28 665
	Purchases	34 000	36 000	40 000	42 000	44 000	44 000
	Paid 1 month (40%)		13 600	14 400	16 000	16 800	17 600
	Paid 2 months (60%)			20 400	21 600	24 000	25 200
W3	Total payments			34 800	37 600	40 800	42 800
W4	Administration and Distribution						
	Total expense		10 000	10 000	10 000	11 000	11 550
	Depreciation		1 000	1 000	1 200	1 200	1 400
	Net expense		9 000	9 000	8 800	9 800	10 150
	Paid same month (80%)			7 200	7 040	7 840	8 120
	Paid 1 month (20%)			1 800	1 800	1 760	1 960
	Total payment			9 000	8 840	9 600	10 080

W5 Dividends paid

Original share capital $600 000 + share issue (face value) (100/150 × $75 000, i.e. $50 000)

Dividend is 10p × 650 000 = $65 000

★ **Exam tips**

Key features of the model answer

➤ A full title has been provided.

➤ Detailed workings are provided for all the calculations required to complete the cash budget.

➤ The receipts from trade receivables takes account of all the information provided.

➤ The share issue has been recorded in the correct month.

➤ The information about payments to trade payables has been correctly interpreted.

➤ Depreciation has been treated as a non-cash expense and deducted from the expense figures.

➤ The dividend payment has been based on the face value of shares.

1   A company's cash budget shows that the business's overdraft limit will be exceeded during several months in the budget period. The directors have considered switching to cheaper suppliers of materials and increasing the selling price of products in order to improve the budgeted cash position. Discuss the likely effectiveness of these ideas.

2   T Limited's statement of financial position at 31 December 2016 was as follows:

Budgeted Statement of Financial Position at 31 December 2017

Assets	$000
Non-current assets	
Property, plant and equipment	480
Current assets	
Inventory	35
Trade receivables	29
Cash and cash equivalents	13
	77
Total assets	557
Equity and liabilities	
Equity	
Ordinary shares of $1 each	300
Share premium	30
Retained earnings	70
	400
Current liabilities	
6% Debentures (2017)	100
Trade payables	36
Taxation	21
	157
Total equity and liabilities	557

The following forecasts have been prepared for the year ending 31 December 2017:

• Sales and purchases for each of the four quarters ending 31 December 2017 (all sales and purchases are on credit)

	Purchases	Sales
	$000	$000
January–March	51 000	99 000
April–June	54 000	108 000
July–September	57 000	114 000
October–December	54 000	105 000

• Purchases are made two months before the month of sale and trade suppliers allow two months' credit. (At 31 December 2016 trade payables are: $18 000 relating to purchases made in November 2016 and $18 000 relating to purchases made in December 2016.)

• Credit customers are allowed one month's credit and are allowed a cash discount of 5 per cent.

• The debentures will be redeemed on 1 July 2017. The redemption will be fully financed by an issue of ordinary shares of $1 each at a premium of 25c per share to be made on 1 June 2017.

• Surplus plant and equipment with a net book value of $40 000 will be sold in September 2017. Sale proceeds are expected to be $45 000. The company does not depreciate non-current assets in the year of sale. Non-current assets are depreciated at 20 per cent per annum using the reducing balance method. There are no plans to purchase additional non-current assets.

• Payments for operating expenses will total $9000 per month.

• Part of the company's property will be sublet commencing on 1 August 2017 at a quarterly rent of $6000 payable in advance.

• A dividend of 20 per cent on ordinary shares will be paid on 31 March 2017.

• The taxation liability will be paid on 30 September 2017. The tax liability for the year ending 31 December 2017 is expected to $5000.

• The company's inventory at 31 December 2017 will be valued at $43 000.

(a) Prepare a cash budget for the year ending 31 December 2017.

(b) Prepare a budgeted income statement of the year ending 31 December 2017.

(c) Prepare a budgeted statement of financial position at 31 December 2017.

Key topics

➤ standard costing and its purposes

➤ setting standards

➤ advantages and disadvantages of standard costing

➤ standard hours

➤ variance analysis

➤ greatest challenges

➤ important techniques.

✓ What you need to know

This unit is concerned with four topics: calculating variances for materials, labour, sales and fixed overheads; reconciling standard cost and actual cost; reconciling standard profit and actual profit; causes of variances and their interrelationships; and advantages and disadvantages of standard costing.

19.1 Standard costing and its purposes

Standard costing is a technique which uses predetermined details of costs and revenues which have been carefully calculated to be achievable and which ought to occur in reasonable conditions. Standard costing enables comparisons to be made between what a product should cost to make and what it actually costs to make, and between the revenues which should arise from the sale of products to revenues that actually arise from the sale of products.

Uses of standard costing

➤ comparing actual costs with standard costs and managers using this information to focus on taking corrective action wherever possible to avoid adverse variances (i.e. management by exception)

➤ providing data for budgeting and decision making

➤ providing data for quoting for jobs

➤ providing costs for inventory valuation

➤ providing targets which are intended to motivate employees.

Ideal v. attainable standards

Ideal standards occur under perfect conditions and ignore the possibility of wastage, machine downtime, etc. These standards are almost always avoided as they would be demotivating for staff who would never be able to achieve them. Attainable standards are based on more realistic conditions, which are usually challenging but achievable, and are far more likely to motivate employees and so benefit the business.

> **Key term**
>
> **Standard costing:** a technique which makes use of expected levels of performance which can then be used as a point of reference for comparisons with actual levels of performance.

> **Key term**
>
> **Management by exception:** making efficient use of management time by focussing on areas which need attention and ignoring those that are running smoothly.

19.2 Setting standards

Direct materials

For each product standards are set for the following:

➤ **Usage**: based on the types of direct materials required, the quality of materials and quantity taking into account likely wastage based on direct labour and machinery to be used in the production processes.

➤ **Price**: based on information from suppliers price lists, trade and cash discounts which will be available, expected changes in prices, economic forecasts about inflation, predicted changes in exchange rates.

Direct labour

For each product standards are set for the following:

➤ **Efficiency**: measures the quantity of direct labour to be used and takes account of the operations required to make one unit and the quality of the direct labour to be used as well as including allowances for breaks, preparation, clearing up etc. .

➤ **Rate**: measures the cost of direct labour and will take account of current and agreed future pay rates, overtime rates.

19.3 Advantages and disadvantages of standard costing

Advantages

➤ Provides useful information which can be used by managers to control a business.

➤ Possibility that staff motivation will be improved because they have realistic and achievable targets.

➤ Provides data which makes the preparation of budgets easier.

Disadvantages

➤ Setting up a standard costing system is likely to be expensive because it is time consuming and often requires the involvement of technical experts.

➤ The system is likely to require regular updating, particularly if sophisticated technology-based processes are used.

➤ Unless the system is kept up to date the standards set may be too high or too low whereupon all the potential advantages of the system will be lost.

19.4 Standard hours

Where a business produces a range of products it is helpful to use the concept of a standard hour which is a measure of the quantity of work which can be undertaken in an hour. Output can be measured in standard hours as in the following example:

Key term

Standard hour: a measure of the quantity of work which should be produced in one hour.

Weekly output in Assembly department	
80 units Product X each taking 15 minutes	20 standard hours
150 units Product Y each taking 20 minutes	50 standard hours
Total standard hours of production	70 standard hours

19.5 **Variance analysis**

➤ Calculating variances alerts managers to any deviation from the plan and is one of the most valuable aspects of standard costing. Variances could be adverse (indicating a negative impact on profits) or favourable (indicating a positive impact on profits).

➤ Managers should always be in a position to investigate and explain any variance.

➤ Variances must be based on flexed budgets to ensure comparisons are made on data for the same level of production.

Interrelationship between cost variances

Cost variances can result from a single cause. For example, an adverse material usage variance could have been caused by using poor quality material resulting in more wastage but also causing a labour efficiency adverse variance because production staff are having to spend more time than usual as result of using the poor-quality material.

19.6 **What are the greatest challenges?**

➤ Remembering the formula for each variance.

➤ Correctly recording the outcome of variance calculations.

➤ Explaining the possible reasons for variances.

➤ Explaining the possible interrelationships between variances.

➤ Reconciling budgeted and actual information (total costs and profits).

19.7 **Review some important techniques**

Calculating variances

Variances are described as favourable or adverse and can be indicated by (F) for favourable or (A) for adverse.

Direct materials

▼ **Table 19.1** Direct materials

Direct materials				
(AP = actual price; AQ = actual quantity; SP = standard price; SQ = Standard quantity)				
	Difference is the TOTAL VARIANCE			
AQ × AP Actual cost of production	Difference is the PRICE variance	AQ × SP What should have been paid for the quantity used	Difference is the USAGE variance	SQ × SP What production should have cost

▼ **Example 19.1** Calculating direct material variances

F Limited budgeted to produce 10 000 units using 30 000 kg of materials costing $120 000. Actual production was of 10 000 units using 3.5 kg per unit costing $3.00 per unit.

Step 1: analyse data to obtain information per unit and information for all production

Standard: unit cost is 3 kg at $4 per kg – production 30 000 kg costing $120 000

Actual: unit cost is 3.5 kg at $3 per kg – production 35 000 kg costing $105 000

Step 2: calculate variances

	Difference is the TOTAL VARIANCE = $15 000 (F)			
AQ × AP $105 000	Difference is the PRICE variance = $35 000 (F)	AQ × SP $140 000	Difference is the USAGE variance = $20 000 (A)	SQ × SP $120 000

★ **Exam tip**

Always check that the price + usage variances = total variance.

Here $35 000 (F) + $20 000 (A) do equal the total variance of $15 000 (F).

Direct labour (including flexing the budget)

Direct labour
(AR = actual rate AH = actual hours; SR = standard rate; SH = Standard hours)

✗ **Common error**

Make sure you don't forget to label each variance – it is usually acceptable to use F or Favourable, A or Adverse. However, do NOT use + sign for favourable and – sign or brackets for adverse.

	Difference is the TOTAL VARIANCE			SH × SR What production should have cost
AH × AR Actual cost of production	Difference is the RATE variance	AH × SR What should have been paid for the quantity used	Difference is the EFFICIENCY variance	

▼ **Example 19.2** Calculating labour variance

G Limited budgeted to produce 500 units using 8 hours per unit at $9.00 per labour hour.

Actual production was 400 units which used 2800 hours and cost $28 000.

Step 1: flexing the budget and analysing data to obtain information per unit and information for all production

The first thing to notice is that actual production was different to standard production, so it is necessary to flex the budget. In this case it is necessary to scale down the budget so that it is based on 400 units.

Standard: unit cost is 8 hr at $9.00 per hour – production 3200 hrs costing $28 800

Actual: unit cost is 7 hr at $10.00 per hour – production 2800 hrs costing $28 000

Step 2: calculate variances

💡 **Remember**

When the budgeted production is different to the actual production, it is the budgeted production which is scaled up or down to equate with actual production.

	Difference is the TOTAL VARIANCE = $800 (F)			
AH × AR $28 000	Difference is the RATE variance = $2 800 (A)	AH × SR $25 200	Difference is the EFFICIENCY variance = $3 600 (F)	SH × SR $28 800

Sales variances

	Sales			
	(AP = actual price AQ = actual quantity; SP = standard price; SQ = standard quantity)			

	Difference is the TOTAL VARIANCE			
AQ × AP Actual total sales	Difference is the PRICE variance	AQ × SP What should actual sales have been	Difference is the VOLUME variance	SQ × SP What total sales should have been

▼ **Example 19.3** Calculating sales variances

H Limited's budgeted sales were 1000 units at a selling price of $8.00 per unit. The actual sales were 900 units at $9.00 per unit.

	Difference is the TOTAL VARIANCE = $100 (F)			
AQ × AP $8 100	Difference is the PRICE variance = $900 (F)	AQ × SP $7 200	Difference is the VOLUME variance = $800 (A)	SQ × SP $8 000

Remember

Sales budgets are **not** flexed.

Fixed overhead variances

	Fixed overheads			
	Difference is the TOTAL FIXED OVERHEAD VARIANCE			What fixed overheads should have been applied to actual production
Actual fixed overheads incurred	Difference is the EXPENDITURE variance	Budgeted fixed overhead (ignoring actual production)	Difference is the VOLUME variance	

▼ **Example 19.4** Calculating fixed overhead variances

J Limited have provided the following information for one month's production:

	Actual	Budget
Production	800 units	1 000
Machine hours	220	250
Fixed overhead costs	$5 300	$5 000

Step 1: it is useful to look carefully at data to get an overview

Without making any calculations it is possible to see that the situation is generally unfavourable: production was far less than expected yet fixed overheads were higher.

Step 2: calculate overheads

Actual fixed overheads incurred $5 300	Difference is the TOTAL FIXED OVERHEAD VARIANCE			What fixed overheads should have been applied to actual production $4 400 (**W1**)
	Difference is the EXPENDITURE variance $300 (A)	Budgeted fixed overhead (ignoring actual production) $5 000	Difference is the VOLUME variance $600 (A)	

W1:

Fixed overhead rate is $5000/250 mch hrs, i.e. $20 per machine hour.

Production is expected to be four units per machine hour (i.e. budget 1000 units/250 mch hrs), so actual production of 800 units should have taken 200 hours.

Fixed overheads that should have been applied to actual production = 220 × $20 = $4400.

Variances are:

Total fixed overhead variance: $900 (A)

Fixed overhead expenditure variance: $300 (A)

Fixed overhead volume variance: $600 (A)

Explaining variances

The following table gives some possible explanations for variances:

Table 19.2 Explaining variances

Variance		Favourable	Adverse
Materials	Price	➤ Lower quality material supplied. ➤ Unforeseen price reduction (improved currency exchange rate). ➤ Larger orders made gaining trade discount.	➤ Higher quality material supplied. ➤ Unforeseen price increase. (inflation rate higher than expected, unfavourable change in exchange rate). ➤ Smaller orders made losing trade discount.
	Usage	➤ Better quality materials available. ➤ Higher skilled workers used. ➤ Unforeseen improvements made to machinery/equipment so less wastage.	➤ Poorer quality of materials used. ➤ Less skilled workers used. ➤ Machinery/equipment faulty.

Labour	Rate	➤ Less skilled workers used – earning lower pay rate. ➤ Less overtime working than expected. ➤ Bonuses not awarded.	➤ Higher skilled workers used – earning higher pay rate. ➤ More overtime working than planned. ➤ Unforeseen increase in pay rates.
	Efficiency	➤ Higher skilled workers used. ➤ Better quality materials available. ➤ Unforeseen improvements made to machinery/equipment so less time taken.	➤ Lower skilled workers used. ➤ Lower quality materials used. ➤ Machinery/equipment faults leading to idle time. ➤ Problems with working conditions leading to poor staff morale.
Sales	Price	➤ Less competition. ➤ Increase costs force increase in prices.	➤ More competition. ➤ Reduced costs enable reduction in prices.
	Volume	➤ More effective marketing. ➤ Less competition. ➤ Improved reputation (better quality products). ➤ Change in buying habits of customers.	➤ More competition. ➤ Less effective marketing. ➤ Decline in reputation (bad publicity). ➤ Change in buying habits of customers.
Fixed overheads	Expenditure	➤ Reduced charges by suppliers.	➤ Increased charges by suppliers.
	Volume	➤ More production than expected.	➤ Less production than expected.

Reconciliation statements for cost and for profit

▼ **Example 19.5** Cost reconciliation

K Limited's total budgeted costs were $27 480. The actual cost of direct materials was $11 520 and the actual cost of direct labour was $14 260 giving total actual costs of $25 780.

Variances were as follows:

	$	
Material price	450	(A)
Material usage	1 320	(F)
Labour rate	390	(A)
Labour efficiency	1 220	(F)

K Limited

Reconciliation of budgeted costs with actual costs

	$	
Total budgeted direct costs	27 480	
Add: material price variance	450	(A)
Less: material usage variance	(1 320)	(F)

> **Remember**
>
> Favourable variances **reduce** the budgeted direct costs because less was spent than expected.

Add: labour rate variance	390	(A)
Less: labour efficiency variance	(1 220)	(F)
Total actual direct costs	25 780	

▼ **Example 19.6** Profit reconciliation

L Limited's budgeted profit was $17 450. The following variances occurred:

	$	
Sales price	1 480	(F)
Sales volume	2 050	(A)
Material price	310	(A)
Material usage	220	(A)
Labour rate	560	(F)
Labour efficiency	490	(A)

L Limited

Reconciliation of budgeted profit with actual profit

	$	
Budgeted profit	17 450	
Add: Sales price	1 480	(F)
Less: Sales volume	(2 050)	(A)
Less: material price variance	(310)	(A)
Less: material usage variance	(220)	(A)
Add: labour rate variance	560	(F)
Less: labour efficiency variance	(490)	(A)
Actual profit	16 420	

> **Remember**
>
> All adverse variances **reduce** the budgeted profit.

Revision checklist

I can:

➤ explain how standards are set ☐

➤ remember and the formula for each variance ☐

➤ apply the variances and describe them correctly ☐

➤ explain the causes of variances ☐

➤ reconcile budgeted costs with actual costs; reconcile budgeted profit with actual profit. ☐

Raise your grade

M Limited operates a standard costing system. The following information is available for December 2016:

Budget	
Produce and sell 9 600 units	
Sales at $32.00 each	$307 200
Direct materials (13 440 kilos at $4.50 per kilo)	$60 480

Actual	
Produce and sell 9 120 units	
Sales at $33.00 each	$300 960
Direct materials (11 856 kilos at $4.70 per kilo)	$55 723.20

Calculate the following variances:

a) sales volume

b) sale price

c) total sales

d) direct materials price

e) direct materials usage

f) total direct materials.

Student answer

Sales variances:

a)   Volume $9600 (F) ❶ ❷
....
b)   Price $15840 (A) ❶ ❷
....
c)   Total $6240 (A) ❷
....
Materials variances

Actual cost: $55 723.20
....
Actual usage × standard price = 11 856 kilos × $4.50 = $53 352 ❸
....
Standard cost: $60 480
....
d)   Price: $55 723.20 (Actual) – $53 352 = – 2371.20 ❸ ❹
....
e)   Usage: Actual kilos at SO $53 352 – Standard cost $60 480 = +7128 ❸ ❹
....
f)   Total: Standard cost $60 480 – $55 723.20 = +4756.80 ❸ ❹
....

How to improve this answer

① The candidate has used the wrong formula (comparing actual revenue and standard revenue with the wrong figure, i.e. standard quantity and actual price).

② There are no workings.

③ The candidate has forgotten to flex the budget and should have scaled down standard costs at 9600 units to standard costs at 9120 units.

④ The candidate has not labelled the variances correctly having omitted the $ sign and used + and – rather than Favourable/(F) and Adverse/(A).

Model answer

Sales variances

Workings:

Actual sales: $300 960

W1: Actual quantity at standard selling price = 9120 × $32 = $291 840

Standard sales: $307 200

a) Volume: Budget sales $307 200 – Actual Q at SP (W1) $291 840 = $15 360 (A)

b) Price: Actual sales $300 960 – Actual Q at SP (W1) $291 840 = $9120 (F)

c) Total: Actual sales $300 960 – Budget sales $307 200 = $6240 (A)

Materials variances

Workings:

Actual cost: $55 723.20

W2: Actual kilos × SP = 11 856 × $4.50 = $53 352

Standard cost: $57 456

d) Price: Actual cost $55 723.20 – AQ × SP (W2) $53 352 = $2371.20 (A)

e) Volume: Standard cost $57 456 – AQ X SP (W2) $53 352 = $4104 (F)

f) Total: Standard cost $57 456 – Actual cost $55 723.20 = $1732.80 (F)

★ Exam tips

Key features of the model answer

➤ Make sure you use the correct formula – here the actual sales quantity at budgeted prices has been used in the sales variances.

➤ Always provide workings to support your answer – it is always helpful to label figures in workings.

➤ Always check to see if the budgeted details should be flexed because actual production was different to budgeted production – here the costs have been scaled down to actual production levels.

➤ Check to make sure that variances are correctly labelled – always use the $ sign for amounts and indicate the type of variance (F) or (A).

Exam-style questions

1 State **three** possible causes of an adverse labour rate variance.

2 Describe how standards would be set for material prices.

3 Explain how a favourable labour efficiency variance could be interrelated with an adverse material price variance.

4 N Limited has provided the following information for one month's production:

	Actual	Budget
Production	990	900
Machine hours	155	150
Fixed overhead costs	$2 700	$2 550

Calculate: (i) total fixed overhead variance, (ii) fixed overhead expenditure variance, (iii) fixed overhead volume variance.

5 P Limited manufactures a single product and uses a system of standard costing.

The following information is available for October 2016.

Budget:

Production and sales	7 200 units
Selling price	$96 per unit
Direct materials	4.8 kilos at $11 per kilo
Direct labour	2.5 hours at $6.40 per hour

Actual results

Production and sales	7 800 units
Selling price	$95 per unit
Direct materials	35 880 kilos costing $401 856
Direct labour	21 840 hours costing $146 328

Required

(a) Explain the term 'ideal standard'.

(b) Calculate the following variances for October 2016:

(i) Sales total variance.

(ii) Sales price variance.

(iii) Sales volume variance.

(iv) Direct materials total variance.

(v) Direct materials price variance.

(vi) Direct materials usage variance.

(vii) Direct labour total variance.

(viii) Direct labour rate variance.

(ix) Direct labour efficiency variance.

(c) Prepare a statement reconciling the total budgeted direct costs with the total actual direct costs for October 2016.

(d) Explain possible reasons for the direct material variances.

(e) Recommend ways in which any adverse labour variances could be improved.

Key topics

➤ investment appraisal

➤ net cash flows

➤ the time value of money

➤ cost of capital

➤ techniques

➤ sensitivity analysis and investment appraisal

➤ greatest challenges

➤ important techniques.

✓ What you need to know

This unit is concerned with three topics: calculating net cash inflows and net cash outflows arising from a project; investment appraisal techniques: their application, advantages and disadvantages; and investment decisions and recommendations using supporting data.

20.1 Investment appraisal

Investment appraisal is an important aspect of decision making because of the large sums of money involved, the long-term commitment of resources, and the implications for the future survival of the business. The techniques involved use information about future net cash flows (and sometimes profitability) to determine whether a capital project (i.e. large-scale capital expenditure) will be worthwhile. It is often necessary to consider the wider implications of a particular project – including possible consequences for the workforce, for the environment and for the local economy.

20.2 Net cash flows

Net cash flows are used in many of the investment appraisal techniques. Net cash flows result from finding the difference between cash inflows (based on estimates of future cash receipts arising from the capital project) and cash outflows (based on estimates of future cash expenditure arising from the capital project). Inevitably future net cash flows may be difficult to predict with any degree of accuracy.

20.3 The time value of money

When assessing the value of a capital project it is important to consider that money received now and is more valuable than money received at some future date because of its potential earning capacity. In other words, $100 received now is more valuable than $100 earned, for example, in five years' time, because the money received now could be earning interest during that time period.

20.4 Cost of capital

Cost of capital is a factor which needs to be taken into consideration if the NPV technique is used in investment appraisal. It is a reference to the

💡 Remember

If required, consider non-financial factors such as the effect on employees, the local community and the environment of a capital investment project.

✗ Common error

When estimating net cash flows, remember not to include all expenses in the calculation, and that that depreciation (as a non-cash item) should be excluded from the figure for expenses.

Key term

Cost of capital: the probable cost of finance which will be applied in a capital investment project.

probable average cost of the finance which will be used to cover the capital expenditure involved. For example, the cost of capital could be based on the weighted average of expected ordinary share dividend rate and the interest rate to be charged on loans raised, if both of these methods are used to finance a project. It is important to recognise that the cost of capital is a prediction and may prove to be inaccurate when a capital project actually goes ahead.

20.5 Techniques

Table 20.1 shows which techniques are covered:

▼ **Table 20.1** Techniques

Payback		
Description of technique	Advantages	Disadvantages
Payback measures the time taken for the future cash inflows to equal the cash outflows. The technique emphasises the importance of getting back the initial outlay for a capital project as soon as possible, so that any loan required can be repaid and surplus funds used in other ways to benefit the business.	➤ The technique is based on a simple idea and so it is easily understood by non-experts. As a result it is widely used. ➤ The technique recognises the value of positive net cash inflows at an early stage in the life of a capital project, reducing the risks involved in an investment.	The technique ignores: ➤ the time value of money* ➤ the benefits arising from net cash inflows after the payback period ➤ that some projects may take time to get established and could be beneficial in the longer-term.
*There is an alternative version of the payback technique which does take account of the time value of money called the discounted payback method (see Example 20.4.)		
Accounting rate of return (ARR)		
ARR measures the return on the average investment over the period of the project. The technique uses profit (not cash flows) and so it has some similarities with the ratio which measures the return on capital employed.	➤ It is relatively simple to calculate and to understand. ➤ The results can be compared to the business's usual profitability. ➤ Uses data which is readily available.	➤ It ignores the time value of money. ➤ It uses profits rather than cash flows, so the choice of depreciation methods can have an impact on results.
Net present value (NPV)		
NPV uses a process which converts the value of all future net cash flows to their present value using a discounting technique: the further into the future the net cash flow the lower its present value. The predicted cost of capital is used to determine the discount factor.	➤ Takes account of the time value of money ensuring comparability of all net cash flows. ➤ Takes account of all net cash flows from the project.	➤ More complex calculations are required. ➤ Relies on the use of cost of capital which may be difficult to predict.

Internal rate of return (IRR)		
The technique is used to find the discount rate at which the net present value is zero. If the result is higher than the cost of capital the project is considered worthwhile.	➤ Takes account of the time value of money. ➤ Takes account of all the net cash flows from the project.	➤ The most complex technique to use. ➤ Requires an element of trial and error to obtain a result. ➤ Requires a comparison with the cost of capital which may be difficult to predict.

20.6 Sensitivity analysis and investment appraisal

Because many of the key factors used in an investment appraisal are difficult to predict (for example, expected net cash flows, cost of capital, etc.) and because the investment may involve very large sums, it is important that the decision makers take account of the degree of risk involved in each aspect of the calculations. A sensitivity analysis would indicate the degree of risk involved taking account of these factors over a range of possible outcomes. For example, the analysis could show that there is a 20 per cent risk (a low risk) that the net cash flows for the third year of the project could be significantly lower than the expected amount, or that there is a 60 per cent risk (a high risk) that the cost of capital for that year could be too high for the project to be adopted.

20.7 What are the greatest challenges?

➤ Calculating net cash flows.

➤ Remembering the methods to be applied in each of the techniques.

➤ Making accurate calculations.

➤ Interpreting the results of an investment appraisal.

➤ Reporting effectively on an investment appraisal.

20.8 Review some important techniques

Net cash flows

▼ **Example 20.1** Calculating net cash flows

The directors of HH Limited have decided to switch production to a new product which will require the purchase of new machinery costing $150000. The cost of making one unit of the new product is expected to be:

	$
Direct materials	5
Direct labour	10
Variable overheads	4
Fixed overheads (depreciation)	2
Total cost per unit	21

Additional information:

➤ Demand has been forecast to be 10 000 units Year 1, increasing year on year by 10 per cent for the following two years when production will cease. The selling price per unit will be $30 per unit in Year 1 and then for $32 per unit in Year 2 and $33 per unit in Year 3.

➤ Production staff will receive training in the use of the new machinery at the beginning of the first year, costing $10 000.

➤ All production costs are expected to remain the same over the three-year period except for direct labour which is likely to increase by 10 per cent in Year 2 and by a further 5 per cent in Year 3.

➤ The machinery will require maintenance; it is expected the annual charge will be $5000 in Year 1 increasing by 20 per cent year in Year 2 and by a further 20 per cent in Year 3.

➤ At the end of the four-year period it is estimated that the machinery will be sold for scrap for $10 000.

The detailed calculations are shown below:

	YEAR 1		YEAR 2		YEAR 3	
		$		$		$
Sales	10 000 × $30	300 000	11 000 × $32	352 000	12 100 × $33	399 300
Disposal of machinery						10 000
Total cash inflows		300 000		352 000		409 300
Direct materials and						
variable overheads	10 000 × $9	90 000	11 000 × $9	99 000	12 100 × $9	108 900
Direct labour	10 000 × $10	100 000	11 000 × 11	121 000	12 100 × $11.55	139 755
Maintenance		5 000		6 000		7 200
Training		10 000				
Total cash outflows		205 000		226 000		255 855
Net cash flows		95 000		126 000		153 445

Payback

▼ **Example 20.2** Using the payback method

Using the results from Example 20.1 it is possible to work out the payback period for the new machinery (cost $150 000).

	Net cash flows (from Illustration 1)	Cumulative cash flows
	$	$
Period 1	95 000	95 000
Period 2	126 000	221 000
Period 3	153 445	374 445

By calculating the cumulative cash flows it is possible to see that the initial investment of $150 000 is recovered at some point in Year 2.

The payback point is reached in Year 2 when a further $55 000 net cash flow is achieved (i.e. cumulative cash flow at the end of Year 1 $95 000 + $55 000 gives the required payback of the initial $150 000).

The exact point in Year 2 can be point by comparing the required extra net cash flow of $55 000 with the total net cash flow for Year 2 ($126 000).

The answer can be expressed in a number of ways:

In years	$1 \text{ yr} + \dfrac{\$55\,000}{\$126\,000}$	i.e. 1.44 years
In years and months	$1 \text{ year} + \dfrac{\$55\,000}{\$126\,000} \times 12$	i.e. 1 year 5.24 months
In years and days	$1 \text{ year} + \dfrac{\$55\,000}{\$126\,000} \times 365$	i.e. 1 year 159.33 days

★ **Exam tip**

If, for example, you are rounding the number of days for a payback period, always round upwards – for example, 87.21 days should become 88 days.

Net present value

▼ **Example 20.3** Using the net present value method

MM Limited will be introducing a new product range. The capital expenditure will be $200 000 and this amount will be paid at the start of the project.

The expected cash flows arising from this project are:

Year	Inflows	Outflows
	$000	$000
1	94	47
2	136	69
3	188	101
4	144	73

It is assumed that the cash flows occur at the end of each year and that the capital expenditure will have nil scrap value.

The directors expect the cost of capital will be 9 per cent. The discount factor (present value of $1) based on this cost of capital is as follows:

	Discount factor
Year 1	0.917
Year 2	0.842
Year 3	0.772
Year 4	0.708

★ **Exam tip**

It is usual to express the answer correct to two decimal places.

💡 **Remember**

Check whether the answer requires an answer in years, years and months or years and days.

The net present value calculation is as follows:

Year	Net cash flow	Discount factor	Present value
	$000		$
0	(200)	1.000	(200 000)
1	47	0.917	43 099
2	67	0.842	56 414
3	87	0.772	67 164
4	71	0.708	50 268
Net present value			16 945

The moment the project begins is described as Year 0.

The discount factor to be applied at Year 0 is 1.000.

The result of the calculation is positive. This should be interpreted as meaning that the project is worth considering since the net present value of the net cash inflows exceeds the cost of the investment. A project with a negative net present value is not worth adopting from a financial point of view.

Discounted payback

The criticism of the payback method that it ignores the time value of money can be overcome if net cash flows are discounted. Where this method is used it is necessary to assume that net cash flows arise evenly throughout the year (unlike the net present value method).

▼ **Example 20.4** Using the discounted payback method

Using the data in Example 20.3 (on the previous page and above) but this time assuming the net cash flows arise evenly throughout the year.

Year	Net cash flow	Discount factor	Net present value	Cumulative net present value
	$000		$	$
1	47	0.917	43 099	43 099
2	67	0.842	56 414	99 513
3	87	0.772	67 164	166 677
4	71	0.708	50 268	216 945

The capital investment of $200 000 is recovered during Year 4. The calculation of the exact point in Year 4 is as follows:

Discounted net cash flow required in Year 4: $200 000 − $166 677 = $33 323

Payback (in months) is achieved: 3 years + $\dfrac{(\$33\,323 \times 12)}{\$50\,268}$ i.e. 3 years 7.95 months.

Accounting rate of return (ARR)

▼ **Example 20.5** Using the accounting rate of return

Q Limited are considering the purchase of a new plant and machinery costing $700 000. The forecast profits for the lifetime of the new plant and machinery are estimated to be as follows:

	$000
Year 1	40
Year 2	80
Year 3	100
Year 4	90
Year 5	50

At the end of Year 5 the scrap value of the plant and machinery has been estimated as $40 000.

The formula used to calculate the accounting rate of return is:

$$\frac{\text{Average annual profit}}{\text{Average investment}} \times 100$$

The average investment is calculated using the following formula:

$$\frac{\text{Initial investment} + \text{scrap value}}{2}$$

The average annual profit ($000s) is: $\dfrac{40 + 80 + 100 + 90 + 50}{5}$ i.e.

$\dfrac{360}{5} = 72$

The average investment ($000s) is: $700 + $40 = 740/2 = 370$

The ARR is: $\dfrac{72}{370} \times 100$ i.e. 19.46%

Internal Rate of Return (IRR)

With this method it is necessary to use a discount factor which gives a positive net present value and another discount factor which gives a negative net present value. These discount factors are chosen randomly but selected to be within a reasonable range near the cost of capital. The following formula is used to determine the IRR:

$$\text{IRR} = \text{Lower discount rate} + \frac{\text{Higher NPV} \times (\text{Higher rate} - \text{Lower rate})}{(\text{Higher NPV} - \text{Lower NPV})} \%$$

If the IRR is higher than the cost of capital then the project should be considered as feasible.

▼ **Example 20.6** Using the internal rate of return (IRR) method

V Limited are proposing to open a new factory. The cost of capital for this project is expected to be 14 per cent. The following information is available:

Discount factor used	Net present value
	$
12%	16 500
21%	(25 500)

Using this information:

➤ The lower rate is 12 per cent.

➤ The higher rate is 21 per cent.

➤ The lower NPV is ($25 500).

➤ The higher NPV is $16 500.

$$\text{IRR} = 12\% + \frac{\$16\,500 \times (21\% - 12\%)}{[\$16\,500 - (\$25\,500)]} \quad \text{i.e.} \quad 12\% + \frac{\$16\,500 \times 9\%}{\$42\,000} \ \%$$

$$\text{IRR} = 12\% + \frac{\$1485}{\$42\,000} \ \%$$

$$\text{IRR} = 15.54\%$$

The IRR is higher (just) than the cost of capital so the project will be considered feasible.

Revision checklist

I can:

➤ explain key terms such as time value of money, cost of capital, etc. ☐

➤ calculate net cash flows ☐

➤ remember and use each of the techniques for investment appraisal (payback, discounted payback, net present value, accounting rate of return, internal rate of return) ☐

➤ make recommendations using supporting data. ☐

 Raise your grade

Q Limited are considering investing in some new machinery to avoid loss of production at one of its factories. The new machinery will cost $120 000 and this will be financed by a loan for this amount. The loan will be received on the first day the machinery is purchased repaid in full at the end of the four-year period. Interest on the loan at 9.6 per cent per annum will be payable throughout the four-year period.

The following information is available about the project:

• Cost of capital: 8 per cent.

• Capital expenditure: $120 000 to be depreciated by 25 per cent per annum using the straight-line method; the machinery is not expected to have any scrap value.

• Revenue in year 1: $90 000.

• Total costs (excluding interest charges) in year 1: $56 000.

It is expected that:

• revenue each year will be 10 per cent higher than the year before

• total costs (excluding interest charges) each year will be 5 per cent higher than the year before.

Discount factor for net present value of $1:

	8%
Year 1	0.926
Year 2	0.857
Year 3	0.794
Year 4	0.735

The directors have decided that if the project goes ahead, plans to improve facilities for staff at the factory will have to be postponed.

(a)   calculate: (i) the net cash flows for each of the years 1–4; (ii) the payback period in years and months, (iii) the net present value

(b)   assess whether the project should go ahead giving reasons for the recommendation.

Student answer

(a)

(i)   Calculation of net cash flows

	Year 1	Year 2	Year 3	Year 4
Inflows	$	$	$	$
Revenue	90 000	99 000	108 900	119 790
Outflows ❶ ❷	(56 000)	(58 800)	(61 740)	(64 827)
Net cash flows	34 000	40 200	47 160	54 963

(ii)   Payback period

	Net cash flow	Cumulative net cash flow
	$	$
Year 1	34 000	34 000
Year 2	40 200	74 200
Year 3	47 160	121 360
Year 4	54 963	176 323

Payback in year 3 + 1 360/121 360 ❸ = 3.01 years ❹

(iii) ❺

0	120 000	1.000	120 000.00 ❻
1	34 000	0.926	31 484.00
2	40 200	0.857	34 451.40
3	47 160	0.794	37 445.04
4	54 963	0.735	40 397.81
		❼	23 778.25

(b) The directors should go ahead with the project because it will pay back the investment within four years and also has a positive net present value. ❽

How to improve this answer

① Total costs have been used for outflows, but these include non-cash depreciation charges.

② Annual interest charges have been overlooked.

③ The payback calculation is year 3 is based on the wrong data.

④ The answer is given in years, but should have been given in years and months.

⑤ The calculation of NPV is not well presented – for example there are no table headings.

⑥ The negative cash flow for the investment should be shown in brackets.

⑦ The final result is not labelled.

⑧ The assessment and recommended is very brief and omits consideration of other factors.

Model answer

(a)

(i)

Calculation of net cash flows	Year 1	Year 2	Year 3	Year 4
	$	$	$	$
Revenue	90 000	99 000	108 900	119 790
Outflows (**W1**)	(31 000)	(33 800)	(36 740)	(39 827)
Interest on loan	(9 600)	(9 600)	(9 600)	(9 600)
Net cash flows	49 400	55 600	62 560	70 363

W1	$	$	$	$
Total costs	56 000	58 800	61 740	64 827
Less depreciation	(25 000)	(25 000)	(25 000)	(25 000)
Cash outflow	31 000	33 800	36 740	39 827

(ii)

Payback period	Net cash flow	Cumulative net cash flow
	$	$
Year 1	49 400	49 400
Year 2	55 600	105 000
Year 3	62 560	167 560
Year 4	70 363	237 923

Payback occurs in year 1 + (50 600/55 600 × 12) months

i.e. 1 year 10.92 months

(iii)

Year	Net cash flow	Discount factor	Net present value
0	(100 000)	1.000	(100 000.00)
1	49 400	0.926	45 744.40
2	55 600	0.857	47 649.20
3	62 560	0.794	49 672.64
4	70 363	0.735	51 716.81
Total net present value			94 783.05

(b) The project has a substantial positive net present value and a payback period well within the time for the loan to be repaid.

However, the directors should also consider:

- whether it will be possible to borrow the sum required for the project

- the impact on the staff of delaying the improvements to their facilities which may lead to a loss of morale

- whether staff retraining will be required so that the new machinery can be used effectively

- whether there are other more viable projects

but also the positive impact on staff of continuing production and therefore providing job security.

Assuming the loan can be secured and staff concerns overcome, it is recommended the project should go ahead.

★ Exam tips

Key features of the model answer

➤ Depreciation has been excluded from the cash outflows and workings provided.

➤ Annual interest charges have been taken into account.

➤ The particular payback time in the second year is based on taking the net cash flow required to reach the total expenditure compared to the net cash flow for the third year.

➤ The payback period is given in the required format.

➤ The net present value calculations are well presented with negative figures in brackets, column headings and a clearly labelled final result.

➤ The assessment includes more detailed points, including both financial considerations such as the likelihood of obtaining the required finance, and non-financial considerations such as staff morale.

? Exam-style questions

1 State the formula used to calculate the accounting rate of return.

2 Describe how a business selects a figure for cost of capital.

3 Explain the treatment of depreciation in investment appraisal techniques.

4 The directors of Y Limited are considering investing in a project which has an initial outlay of $1.2m. The project is expected to lose its value evenly over a five-year period and then be scrapped. The company's cost of capital is expected to be 11 per cent.

The profit at the end of the first year of the project is expected to be $120 000 rising by 15 per cent in year 2 and then by a further 10 per cent in Year 3 and a further 10 per cent in Year 4. In year 5 profits are expected to be 80 per cent of the figure for Year 4. Depreciation will be charged using the straight-line method.

Extracts from present value tables for $1.

	11%	18%
Year 1	0.901	0.847
Year 2	0.812	0.718
Year 3	0.732	0.609
Year 4	0.650	0.516
Year 5	0.594	0.437

(a) Calculate:

  (i)   the accounting rate of return

  (ii) the discounted payback period

  (iii) the net present value

  (iv) the internal rate of return.

(b) Identify **four** factors other than the results of the calculations which might affect a decision whether or not to go ahead with the project.

5 The directors of N Limited have an opportunity to manufacture a new product. This would require the purchase of new equipment costing $140 000 which would be scrapped after four years.

The following forecasts have been made:

- Sales: 6000 units in Year 1 increasing by 20 per cent in Year 2 and by a further 5 per cent in year 3, but then falling to 4000 units in Year 4.

- Selling price: will be fixed at $30 per unit for the first two years, but then increased by 5 per cent year on year thereafter.

- Variable costs: $20 per unit for Year 1, increasing by 5 per cent in Year 2 and then remaining unchanged for Years 3 and 4.

- Fixed costs (excluding depreciation): $8000 per annum.

- Cost of capital: 12 per cent.

Present value table for $1 (extract)

	12%
Year 1	0.893
Year 2	0.797
Year 3	0.712
Year 4	0.636

(a) Calculate the net cash flows for each year.

(b) Calculate the payback period in years and days.

(c) Calculate the net present value.

(d) Advise the directors whether they should proceed with the new product on purely financial grounds. Justify your answer.

21 Raising your achievement

During your course of study

Let's start with some ideas about how to make good use of your study time.

Recent research shows that there are some important things you can do from the very start to make sure you achieve your best. The first thing is to set yourself some definite goals. Here are some important questions which may help you to think a little differently about how to approach your work:

Question 1

What grade do you want to achieve in accounting in order to feel pleased with your result and to ensure that you can gain acceptance for the next stage in your education or employment?

Question 2

How much time do you spend out of class working on developing your skills in accounting? Maybe you already spend a good many hours in private study, but if this is not the case, it is a good idea to increase your study time. This can be done gradually, for example you could plan to add an extra hour to your normal study time for the next few weeks. Then you could add an extra hour to this increased time, and so on.

Question 3

How do you feel when you have received back some work in accounting and you have not done quite as well as you had hoped, despite having made a real effort to get a good mark? You may feel a bit deflated and hope you can get over the disappointment quickly. However, you could think a little differently. Students who say to themselves something like, 'Okay, this did not go so well; let me have a look at what I did wrong and learn from my mistakes' are able to turn a negative experience into something far more positive, which then leads to real progress.

Question 4

What do you do with all the work you complete during your course of study? Maybe your answer is, 'I have it all here somewhere, it just needs sorting out'. Many students would find it helpful if they kept a carefully-organised collection of the work they do. Have a separate section for each topic with your notes and all the answers you have prepared together with the questions. When preparing for the examination, having a well-organised file will mean you can get on with revision without wasting time trying to find the relevant materials.

At this stage you could be working towards the following position:

➤ you have a definite goal (grade you would like to achieve)

➤ you have a plan to increase the time spent on private study

➤ you view each result you achieve for the work you do as an opportunity to learn from your mistakes

➤ you are building up a well-organised file of work which will be invaluable for revision.

★ Exam tip

Setting a specific goal will motivate you and give you something to focus on.

★ Exam tip

Make definite plans to increase your study time.

★ Exam tip

Don't dwell on disappointment, but try to learn from your mistakes.

★ Exam tip

Organising your work will make it easier to revise.

Practical ideas to help you learn more effectively

Self-assessment: this means that you mark your own work making use of a model answer or mark scheme provided with past examination papers. Research has shown that there are many benefits to doing this if you carry out the process on a regular and frequent basis, for example:

➤ you find out for yourself how well you have done after completing a task

➤ you can get immediate information about any errors or omissions in your answer, so you can start to learn from your mistakes right away

➤ if you use mark schemes (for example exam paper mark schemes) you can learn a lot about how marks are allocated to answers so that you are better-prepared to produce the answers expected in an exam situation.

'Repairing' your answers: the more substantial benefits of checking your own work arise when you then go on to look more closely at any aspect of your answer which was not correct or where something was missing. It is a good idea to spend a little time trying to understand why the model answer is showing a particular outcome which is different to the answer you have provided. Try and work out for yourself how the right answer was achieved, or maybe ask your teacher or a friend to explain the right answer to you if you really cannot see how it was done. Make a point of adding some notes about the correct answer to your work; these notes will be useful when you look back at the question and the answer.

Here are some examples of what is meant by repairing answers:

Example 1: repairing an answer where a calculation is incorrect

The question gave the following details about income from rent received:

	$
Opening balance owing	790
Receipts during year	11 940
Closing balance received in advance	500

Comparing the student's answer with the model answer:

Student's answer			Model answer		
	$				$
Opening balance owing	790		Opening balance owing		(790)
Rent received during year	11 940		Rent received during year		11 940
Closing balance received in advance	(500)		Closing balance received in advance		(500)
	12 230				10 650

Action taken to repair the answer:

Repaired answer		
	$	I should have deducted this, as it is income for the previous year.
Opening balance owing	(790)	
Rent received during year	11 940	
Closing balance received in advance	(500)	
	(12 230)	This figure becomes $10 650

Example 2: repairing an answer requiring a written response

The question asked for comments about a decision to switch production away from one particular product to another more profitable product.

Comparing extract from the student's answer with extract from the model answer:

Student's answer	Model answer
Reasons for going ahead with idea:	Reasons for going ahead with idea:
➤ Will possibly make more profit.	➤ Opportunity to make more profit.
Reasons against idea:	Reasons against idea:
➤ Is there a demand for the new product?	➤ Is there a market for the new product?
➤ Will existing machinery be suitable for production?	➤ Will sales volume be sufficient to increase profit?
	➤ Do staff have the skills required to make the new product?
	➤ How will staff react to this change?
	➤ Will staff need retraining? How much would this cost?
	➤ Will machinery need to be adapted or even replaced? How much would this cost?

Action taken to repair the answer:

Repaired answer
Reasons for going ahead with idea:
➤ Will possibly make more profit.
Reasons against idea:
➤ Is there a demand for the new product?
➤ Will existing machinery be suitable for production?

I should have included:

'sufficient demand to make more profit';

a reference to the cost of making changes to machinery.

I forgot to include important points about staff:

their reaction, have they the skills, will they need retraining, cost of retraining.

So why is assessing my own work and then 'repairing' answers such a good idea?

Much international research has shown that this approach, carried out systematically, will make a big difference to what an individual can gain from all the practical work which is carried out, and over a course of study can boost an individual's performance by as much as two grades!

In the first example the student has focussed attention on how to make adjustments at the beginning of the year. If this process is repeated whenever this type of error is made, then very soon it is likely the student will have learned the correct version of this important technique. In the second example, the student has focussed attention on some omissions from the answer. They now have their own personal record of a much better answer, and if they repeat this process for similar questions it is much more likely they will remember what to think about. Remember each correction you make represents a step towards improving your performance.

Making progress with written answers

Many students prefer using accounting techniques to writing about them. In order to make progress on demonstrating knowledge and understanding of key accounting ideas, you could try the following process in a situation where you are reading through a passage in a textbook or handout about a particular topic. The process helps you organise information and consists of several steps:

Step 1: highlight key points in the text.

Step 2: prepare a table in which you make some notes which summarise key points about a particular aspect of the topic.

Step 3: prepare a second table this time making some notes which summarise key points about a different aspect of the topic ... and so on – more tables can be added if necessary.

Example 3: using tables to analyse knowledge and ideas in an AS level topic: share issue

Having highlighted key points in a textbook about bonus issues of shares and rights issues of shares, you could produce tables along the following lines:

Table 1: practical aspects

	Bonus issue	Rights issue
Effect on issued capital	Increases	Increases
Effect on reserves	Decreases	Increase possibly in share premium
Effect on cash	None	Increase in cash

Table 2: implications for shareholders

	Bonus issue	Rights issue
Possible benefits	➤ Gain extra shares. ➤ In future may receive more in dividends.	➤ Chance to buy shares at an advantageous price. ➤ If issue taken up no change in each shareholder's control within the company.
Possible drawbacks	➤ Share price may decrease when issue made, so little or no gain. ➤ Bonus issue may be alternative to cash dividend as company short of cash.	➤ If rights issue not taken up completely by existing shareholders, control could change as new shareholders now own shares.

... and so on with tables perhaps looking at the two types of share issue from the point of view of the company, etc.

The benefit of use this type of approach is that it makes information more accessible because the focus is on one particular aspect of the topic at a time. As a result, the ideas become more understandable and more memorable because they are separated out and carefully organised.

Here is a further illustration for an A Level topic.

Example 4: using tables to analyse knowledge and ideas in an A level topic: computerised accounting systems

You have just been reading some passages in a textbook about the introduction of computerised accounting systems.

Step 1: highlight each key point.

Step 2: set up a table to organise some key information about the topic focussing on just one aspect.

Step 3: set up a second table to organise some key information about the topic focussing on a different aspect ... and so on more tables can be added if necessary.

Table 1: why convert to a computerised accounting system?

	Benefits	Drawbacks
From the business's viewpoint	➤ greater accuracy ➤ greater speed ➤ simultaneous updating of records ➤ improved accessibility ➤ more information available ➤ possible reduction in staffing costs.	➤ initial outlay on equipment etc. ➤ training costs ➤ risk of data loss ➤ cost of maintenance and updating ➤ additional costs during change from manual to computerised system.
From an employee's viewpoint	➤ acquire new and marketable skills ➤ possibility of increase in pay.	➤ may be job losses ➤ may not be able to adapt easily to new systems ➤ may not cope well with training ➤ health issue concerns (eye strain etc.).

Table 2: converting from manual to computerised system

	Problems	Solutions
Changing the system	➤ requires careful planning based on immediate and future needs ➤ making sure there is no loss of information or efficiency.	➤ choose system and software to use ➤ run the manual system alongside the computerised system during a trial/introductory period.
Security issues	➤ loss of data ➤ who has access? ➤ avoiding information getting into the wrong hands.	➤ need for frequent back-ups, possibly automated back-ups ➤ use of passwords ➤ encryption for highly-sensitive information.

Again, the process of analysing text and organising the key points into tables which have a clear focus can act as a powerful aid to learning.

Avoiding the most common weaknesses in examination answers

When you look at the 'Raise your grade' feature in the units in this book you will soon become aware that one common fault in almost every topic is the lack of workings to support calculations made when using accounting techniques. The following illustrates the importance of providing workings where complex calculations are required:

Example 5: the importance of workings in an AS answer

Students were answering a question where a business had not maintained a full accounting system. They had to work out the depreciation charge on the business's equipment making use of the following information:

	At 1 January 2016	At 31 December 2016
	$	$
Equipment net book value	23 900	24 700

The question included details about the sale of some equipment which had a net book value of $1700 for $2300, and the purchase of additional equipment for $4000.

Student's answer **no** workings showing the mark awarded	Student's answer **with** workings showing the mark awarded			Correct answer with mark scheme		
		$			$	
	Opening bal	23 900		Opening bal	23 900	
Depreciation:	Add new	4 000	(1)	Add new	4 000	(1)
$900 (x)	Less disposal	(2 300)	(x)	Less disposal	(1 700)	(1)
	Less closing bal	(24 700)		Less closing bal	(24 700)	
	Depreciation	900	(1 of)	Depreciation	1 500	(1 of)
Total 0/3 marks	*Total 2/3 marks*					

Both students have made the same mistake: they have used the wrong figure in the calculation for the value of the disposal. The first student has shown no workings and, therefore, provided no idea how the answer was obtained and loses all the marks: maybe it was the wrong disposal figure, maybe it was an arithmetical error – who knows? The second student has provided workings, so it is easy to see where the mistake was made and so loses just the one mark for the wrong disposal figure. (By the way '1 of' means 1 mark for own figure – it is usual to award a mark for wrong final figure when a mark has already been lost for a mistake within a calculation. In other words, an effort is always made to ensure an answer is not penalised twice for one mistake. 'Own figure' marks are not given where there are no workings supporting the figure.)

Imagine how the loss of marks could build up in an exam paper if workings are not shown!

A further point about workings

Having established the importance of workings, here is another recommendation about how to show workings in answers: label each figure in the calculation rather than providing just a string of figures. The benefit is that it is so much easier to understand a calculation when each item is clearly labelled. This is true of a correct calculation; it is even more true when there is a slip up in the calculation and someone is trying to award marks to what is right. Of course, in the pressure of the exam room, if time is running short you may, understandably, be forced to cut corners and revert to showing a string of figures just to save some precious time.

Example 6: the best way of showing workings – calculating income from subscriptions in a question about a club (A Level topic)

Students were asked to calculate subscription income from the following information:

	At 1 January 2016	At 31 December 2016
	$	$
Subscriptions owing	700	900
Subscriptions received in advance	400	200

Subscriptions received during the year totalled $15 900.

Workings version 1	Workings version 2	
$15900 – 700 + 900 + 400 – 200 = $16300		$
	Opening balance owing	(700)
	Opening balance in advance	400
	Receipts	15900
	Closing balance owing	900
	Closing balance in advance	(200)
	Income	16300
Comment: the correct answer, but the unlabelled figures are difficult to follow.	*Comment:* when the figures are labelled, it is much easier to understand what is being done.	

You may wonder why the absence of detailed workings is such a common problem, when it is generally quite well known that workings are important. One possible reason is simply that the habit of showing detailed workings was never really developed during the course of study, with individual students promising themselves to make the effort to show full workings when it came to the examination. Unfortunately, this good intention probably did not work out, simply because the students were not used to providing these details and had to abandon the idea under the pressure of the exam.

One other common error

The other problem which arises so frequently is that answers are not well presented. To some extent this is understandable under exam conditions and, of course, it is accepted that quick corrections and alterations may need to be made to answers. However, there is no real reason why financial statements should not have proper headings and be set out properly. Sometimes poorly presented answers arise because the habit of producing financial statements in their correct from has not been adopted.

So, the important idea arising from this review of the two most common weaknesses in exam answers is to **get into the habit of providing detailed workings and presenting answers well from the very start of your course.**

Understanding what is expected by different types of question

Each task in an exam will start with a particular word which asks you to do something. It is important to understand why particular words are used and what they mean, otherwise it is possible you will waste time by misunderstanding what is required and lose marks for not doing what is expected of you.

Examination papers will be designed:

To test	By asking you to
Knowledge	Identify State
Understanding	Explain Compare Distinguish Between
Ability to use accounting techniques in particular situations	Calculate Prepare
Investigate a situation or problem	Analyse Assess
Ability to look at situation from different points of view; consider a range of factors; make a judgement or decision	Discuss Justify Advise Evaluate Recommend

Words used	What is expected		Examples of questions
Identify State	A brief response, enough to show you can remember some facts or ideas.	AS level	**State** the advantages of using marginal costing. **Identify** the margin of safety in the break-even chart.
		A Level	**State** the formula for calculating the price/earnings (P/E) ratio. **Identify** two disadvantages of using the accounting rate of return (ARR) method of investment appraisal.
Explain Compare Distinguish between	A fuller written statement giving facts and (especially where explain is used) requiring some development; that is a further statement designed to show understanding. Sometimes the development can take the form of examples.	AS level	**Explain** the importance of the liquid ratio. **Compare** and contrast the two methods of depreciation. **Distinguish** between ordinary shares and debenture loans.
		A Level	**Explain** how the direct materials variances may have arisen. **Compare** and contrast activity-based costing with absorption costing. **Distinguish between** a cost pool and a cost driver.

Calculate Prepare	Financial statements and/or calculations are required using information about a situation in a particular business.	AS level	**Calculate** the contribution for each product. **Prepare** the income statement for the year ended 31 December 2016.
		A Level	**Calculate** the payback period. **Prepare** the accounts to record the joint venture in the books of each of the participants.
Analyse Assess	Where you study some information in some depth, separating out different aspects of the information.	AS level	**Analyse** the business's performance for the year ended 31 December 2016 in regard to (i) profitability, (ii) liquidity. **Assess** the effect on the partners of the proposed changes to the partnership agreement.
		A Level	**Analyse** the effect on the budgeted profit of the proposed changes. **Assess** the effect on the workforce of the directors' decision.
Discuss Justify Advise Evaluate Recommend	Where you look at both sides of a particular proposal or idea pointing out the benefits but also the drawbacks. In some situations it will be important to support comments with calculations and provide a thorough study of data. Often you will need to give consideration to financial and non-financial considerations. You may be asked to make a judgement and give reasons to support it.	AS level	**Discuss** whether or not the company should continue to make all the products. **Justify** your answer. **Advise** the partners whether they should take out a bank loan or admit a new partner. **Evaluate** whether or not X should have joined the partnership. **Justify** your answer.
		A Level	**Recommend** whether or not the company should continue to use absorption costing. **Justify** your recommendation. **Advise** the directors whether or not the company should take over the partnership. **Justify** your answer by analysing two benefits and two limitations to the company. **Discuss** the factors that should be considered before the special order is considered. **Evaluate** the proposal to purchase the new machinery. Support your answer with calculations.

Revision

As the examinations draw nearer you will be thinking about the best way to revise. Just how much revision is done and how frequently it is done will, of course, vary from one student to another. In accounting it is important that a substantial portion of your revision time is spent actually answering questions. This will ensure that you are skilled in carrying out the accounting techniques, or that you remember the key points to be made in a variety of questions requiring a prose response. You could find it useful to make a checklist of all the main topics in the AS/A level specification and work your way through the list finding questions to try on each topic. This is where that well-organised course file will prove invaluable because you should easily be able to find suitable questions with your answers, including your own notes on points you learned when you did the question originally, plus a model answer. If you feel confident about a topic just try the trickier parts of the question again. Avoid looking at the model answer unless you really get stuck. Don't overlook answering some questions requiring a purely written response, as well as those requiring accounts, calculations, financial statements, etc. If you are less sure about a topic, be prepared to spend more time on it. Try working through the entire question rather than just a few selected elements. Again, have a model answer available to help you if you get stuck. In the end, you will find that actually answering questions will restore your confidence far more effectively than just reading through notes and looking at answers to past questions.

In the examination

There are three important things to keep in mind:

1. Read the question carefully – as you know, accounting questions often contain a great deal of information and it is so easy to overlook some detail or particular wording which can drastically effect your answer.

2. Make sure you have read the key words in each task (identify, prepare, discuss, etc.) so that you know what you are expected to do.

3. Time management (see below).

AS examination

In paper 2 (structured questions) you will have 1 hour 30 minutes to answer 2 question of 30 marks each and 2 questions of 15 marks each, so the time allocation for each question would be:

30 mark question	30 minutes each question
15 mark question	15 minutes each question

A Level examination

In Paper 3 (structured questions) you will have 3 hours to answer 4 questions of 25 marks each, so the time allocation for each question should work out at 45 minutes.

These timings per question are a guide only. You may need to spend a little more time on a particular question and a little less on another. However, don't forget that every minute over the target time will be a minute less available on another question. One thing to avoid is wasting time trying to get something to work out, when that time could be better used on another task. For example, it might be tempting in an AS question to spend time trying to get the totals of a statement of financial position to agree. This could take a while and you might find that it was just a careless error somewhere which would have lost you one mark. All that time spent just to recover one mark!

AS Level

A Level

AS Level / A Level

22 Exam-style questions

Here is a selection of questions similar to those you will be asked in the exams. They are grouped by paper.

Paper 1 multiple choice

Paper 1 consists of 30 multiple-choice questions, based on the AS Level syllabus content.

1 Which of the following should be shown as a credit balance in a trial balance?

   A carriage charges   **C** inventory

   B discounts allowed   **D** returns outwards

2 A bookkeeper has prepared a purchases ledger control account. Which of the following items should not have been included in this account?

   A cash purchases   **C** refund by a credit supplier

   B discounts received   **D** returns outwards

3 Which of the following transactions should be recorded in the general journal?

   1 The purchase of a non-current asset on credit.

   2 The withdrawal of goods by the owner for private use.

   3 The issue of bonus shares by a limited company.

   4 A set-off between a trade receivable and a trade payable.

   A 1, 2, 3 and 4

   B 1, 2 and 3 only

   C 1 and 3 only

   D 2 and 4 only

4 A bank statement shows an overdrawn balance of $175. However, there is an unpresented cheque for $84 and a dishonoured cheque of $118 has been incorrectly recorded as $181 in the bank statement.

   What is the correct bank statement balance?

   A overdrawn $28   **C** overdrawn $196

   B overdrawn $154   **D** overdrawn $322

5 A furniture retailer had the following transactions.

   1 Repayment of bank loan.

   2 Payment to supplier of goods for resale.

   3 Payment of carriage charges on new equipment.

   Which transactions are revenue expenditure?

   A 1 only   **C** 2 only

   B 1 and 3   **D** 2 and 3

6 A trader purchased some equipment costing $20 000 on 1 January 2015. The equipment has been depreciated by 25 per cent per annum using the reducing balance method at the end of each financial year 31 December 2015 and 31 December 2016. The equipment should have been depreciated using the straight-line method by 20 per cent per annum.

Which of the following entries will correct this mistake?

	debit	credit
A	income statement $750	provision for depreciation $750
B	provision for depreciation $750	income statement $750
C	income statement $1250	provision for depreciation $1250
D	provision for depreciation $1250	income statement $1250

7 The owner of a business has been told that some of the business's office furniture is now considered very desirable and could be worth a lot more than the book value. The owner had been advised not to change the value of the office furniture.

Which accounting concept should be applied?

A business entity   C materiality

B going concern   D money measurement

8 The owner of a business has decided to create a provision for doubtful debts.

Which two accounting concepts are being applied?

A going concern   business entity

B materiality   consistency

C prudence   accruals

D realisation   duality

9 On 1 January 2016 rent of $480 was prepaid. During the year rent paid totalled $10 900. On 31 December 2016 rent of $640 was owing.

What amount should be charged for rent in the income statement for the year ended 31 December 2016?

A $9 780   C $11 060

B $10 740   D $12 020

10 On 1 January 2016 a business had a provision for doubtful debts of $540. At 31 December 2016 trade receivables totalled $9 800. It was decided to maintain the provision for doubtful debts at 5 per cent of trade receivables.

What will be the effect of updating the provision for doubtful debts?

	Profit for the year ended 31 December 2016	Current assets at 31 December 2016
A	Decrease $490	Decrease $50
B	Decrease $490	Decrease $490
C	Increase $50	Decrease $50
D	Increase $50	Decrease $490

11 The following information is available for entry in a statement of financial position.

	$
Expenses accrued	900
Expenses prepaid	500
Income accrued	600
Income prepaid	400

What amount should be included in the current assets?

A $900   C $1300

B $1100   D $1500

12 The following information is available for a sole trader's business for 2016.

	$
Capital, 1 January	37 800
Capital, 31 December	36 500
Drawings	11 300
Additional investment by owner	10 500

What was the profit or loss for the year?

A loss $500   C profit $500

B loss $2300   D profit $2300

13 A trader has not maintained a complete accounting system. The following information is available concerning trade payables for the year ended 31 December 2016.

	$
Opening balance	20 000
Amounts paid to credit suppliers	102 000
Interest charged on overdue accounts	2 000
Returns outwards	8 000
Closing balance	15 000

What was the value of the business's credit purchases for 2016?

A $91 000   C $103 000

B $101 000   D $113 000

14 Which of the following correctly lists the items recorded in a partnership's appropriation account?

A drawings, interest on capital, partnership salaries

B interest on capital, partnership salaries, shares of residual loss

C interest on drawings, interest on a partner's loan, shares of residual profit

D partnership salaries, interest on a partner's loan, interest on drawings

15 The current account of a partner recorded the following:

	$
Opening debit balance	3 000
Partner's salary	24 000
Drawings	11 000
Interest on drawings	1 000
Share of residual loss	5 000

What was the closing balance of the partner's current account?

A $4000 credit   **C** $14 000 credit

B $10 000 credit   **D** $16 000 credit

16 P, Q and R are in partnership sharing profit and losses in the ratio 3 : 2 : 1. R is due to retire from the business after which P and Q will remain in partnership sharing profits and losses equally. It has been agreed that tangible assets will be revalued upwards by $48 000, goodwill will be valued at $36 000 and that no account for goodwill will be maintained in the books of account.

What will be the net effect of these arrangements on the capital account of Q?

A Credit $10 000   **C** Debit $10 000

B Credit $18 000   **D** Debit $18 000

17 Which of the following should not be recorded in a statement of changes in equity?

A bonus issue of ordinary shares

B proposed dividend on ordinary shares

C reserve arising on revaluation of property

D transfer to general reserve

18 A limited company has an authorised share capital of 800 000 shares of $1 each. The directors of a company have announced a dividend of 10 cents per share. The following information is also available.

	$
Issued share capital	400 000
Share premium	100 000
Retained earnings	300 000

What is the total dividend payment?

A $40 000   **C** $70 000

B $50 000   **D** $80 000

19 Which of the following should be included in the equity section of a limited company's statement of financial position?

1 Bank loan

2 Debentures

3 Issued ordinary share capital

4 Share premium

A 1, 2, 3 and 4

B 1, 2 and 3 only

C 1 and 2 only

D 3 and 4 only

20 A partnership's statement of financial position includes the following details:

	$
bank overdraft	5200
inventory	4500
trade and other payables	8800
trade and other receivables	7400

What is the liquid ratio?

A 0.53 : 1   **C** 1.43 : 1

B 0.85 : 1   **D** 1.93 : 1

21 Which of the following would cause a decrease in the current ratio?

A owner taking inventory for own use

B paying trade suppliers

C purchasing additional inventory

D trade receivables delaying payment

22 Which of the following is the correct ratio for calculating trade receivables turnover?

A $\dfrac{\text{trade receivables}}{\text{credit sales}} \times 100$ **C** $\dfrac{\text{credit sales}}{\text{trade receivables}} \times 100$

B $\dfrac{\text{trade receivables}}{\text{credit sales}} \times 365$ **D** $\dfrac{\text{credit sales}}{\text{trade receivables}} \times 365$

23 An increase in which ratio indicates a poorer performance for a business?

A expenses to revenue

B non-current asset turnover

C rate of inventory turnover

D return on capital employed

24 The following information is available about raw materials:

Opening inventory	500 units at $5 each
Purchases	200 units at $6 each
Issued to production	300 units

The business uses the first in first out (FIFO) method of inventory valuation.

What is the value of the closing inventory?

A $1500   **C** $2000

B $1800   **D** $2200

25 A business uses absorption costing. The assembly department is forecast to use 1600 machine hours per month and 2200 labour hours per month.

Which formula should be used to calculate the overhead absorption rate?

A $\dfrac{\text{labour hours}}{\text{total overhead costs}}$ **C** $\dfrac{\text{total overhead costs}}{\text{labour hours}}$

B $\dfrac{\text{machine hours}}{\text{total overhead costs}}$ **D** $\dfrac{\text{total overhead costs}}{\text{machine hours}}$

26 The following costs were incurred by a restaurant:

1   Food used in meals

2   Lighting and heating for kitchen

3   Wages of staff preparing meals

4   Depreciation of kitchen equipment

Which of these are direct costs?

A 1, 2, 3 and 4   **C** 1 and 3 only

B 1, 2 and 3 only   **D** 2 and 4 only

27 A company absorbs overheads using labour hours. The following information is available.

	Overheads	Labour hours
Budgeted	$60 000	20 000
Actual	$52 000	16 000

What was the over absorption or under absorption of overheads?

A $4000 over absorbed   **C** $8000 over absorbed

B $4000 under absorbed   **D** $8000 under absorbed

28 A business sold 2000 units at $30 each. It has fixed costs of $10 000. Production costs per unit were as follows:

	$
Direct materials	4
Direct labour	8
Other direct	2
Variable overheads	3

What was the total contribution?

A $16 000   **C** $32 000

B $26 000   **D** $36 000

29 A business's break-even point increased recently.

Which of the following would have had to be decreased in order for this change to occur?

A fixed costs   **C** selling price

B sales   **D** variable costs

30 A business has provided the following information.

	$
Total fixed costs	30 000
Target profit	45 000
Variable cost per unit	10
Contribution per unit	20

How many units should it sell in order to achieve the target profit?

A 1500   **C** 2500

B 2250   **D** 3750

Paper 2 structured questions

Paper 2 consists of four structured questions, based on the AS Level syllabus content. Questions 1, 2 and 3 are on financial accounting; and question 4 is on cost and management accounting. The paper is worth a total of 90 marks.

1   Youssef, a trader, does not keep accounts on a double-entry basis.

(a) State **three** reasons why Youssef may have chosen not to keep accounts on a double-entry basis.   [3]

He is able to provide the following information about the year ended 31 December 2016.

	1 January 2016	31 December 2016
	$	$
Delivery vehicle, net book value	16 400	14 200
Furniture and equipment, net book value	7 900	12 000
Inventory	11 220	14 900
Trade receivables	5 910	7 070
Trade payables	4 470	5 290
Bank overdraft	1 460	?
General expenses owing	430	
General expenses prepaid		150
Rent prepaid	320	-

(b) Calculate the business's capital at 1 January 2016.   [3]

Additional information

1   Summary of business's bank account:

Receipts	$
Cash sales banked	121 000
Receipts from trade receivables	53 280
Bank loan	8 400
Proceeds from sale of unwanted equipment	480
Payments	$
Purchases of goods for resale	39 740
Trade payables	93 490
Purchase of new equipment	6 600
Fitting of new shelves in delivery vehicle	1 810
General expenses	5 420
Rent	8 400
Wages	23 390

2   Cash sales were all banked with the exception of $1200 per month which Youssef withdrew for personal use.

3   Cash discounts of $490 were received from trade payables.

4   The account of a credit customer who owed $360 had been written off as irrecoverable.

5 All goods were sold with a mark-up of 50 per cent.

6 Some inventory was damaged when the storerooms were flooded early in December 2016. The inventory was not covered by insurance.

7 Unwanted equipment was sold at a loss of $320.

8 The bank loan is repayable in 2021. Interest for the year ended 31 December 2016 of $450 has been included in general expenses.

(c) Prepare the business's income statement of the year ended 31 December 2016. [15]

Youssef has been concerned about the business's profitability for the year ended 31 December 2016.

He has provided the following information about the year ended 31 December 2015:

Profit margin	11.95%
Return on capital employed	33.17%

(d) Calculate the following ratios for the year ended 31 December 2016

   (i) Profit margin [2]

   (ii) Return on capital employed [1]

(e) Evaluate the business's profitability for the year 31 December 2016 compared to the year ended 31 December 2015. Discuss some recommendations to overcome any weaknesses. [6]

[Total: 30]

2 When Yasmin prepared her business's trial balance on the 31 December 2016 the totals did not agree. The difference was entered in a suspense account. The following errors have been discovered.

1 No entries had been made for goods, cost $180, taken for her own use by Yasmin.

2 Discounts received from a supplier, K Patel, of $44 had been correctly recorded in the cash book but credited to the account of a customer, M Patel.

3 Carriage inwards of $112 had been entered on the debit side of the carriage outwards account.

4 Rent received of $360 had been posted in error to the debit side of the administration expenses account.

(a) Describe the difference between an error of principle and an error of commission. [2]

(b) Prepare journal entries to correct errors 1–4. [6]

(c) Make any entries required in the suspense account including the original difference in the trial balance totals. [3]

The draft profit for the year ended 31 December 2016 was $48 727.

(d) Calculate the amended profit for the year ended 31 December 2016. [4]

[Total: 15]

3 The directors of J Limited have decided to make a bonus issue of shares.

(a) State **two** ways in which a bonus issue of shares differs from a rights issue of shares. [2]

The following trial balance was extracted from the books of J Limited after the income statement for the year ended 31 December 2016 had been prepared and before the bonus issue had been made.

	Dr	Cr
	$000	$000
8% Debentures		100
Cash and cash equivalents	24	
Dividends paid	30	
Inventory	27	
Issued share capital: 800 000 shares of 50 cents each		400
Profit for the year		84
Property, plant and equipment at net book value	1 140	
Retained earnings at 1 January 2016		390
Revaluation reserve		120
Share premium		100
Trade and other payables		49
Trade and other receivables	22	
	1 243	1 243

Additional information

A bonus issue of one share for every share held was made on 31 December 2016. The directors wish reserves to be in their most flexible form.

(b) Prepare entries to record the bonus issue in the general journal. [4]

(c) Prepare a statement of changes in equity at 31 December 2016 after the bonus issued had been made. [5]

(d) Prepare an extract from the company's statement of financial position at 31 December 2016 to show the company's equity and liabilities at this date. [4]

[Total: 15]

4 V Limited manufacture a single product which has a selling price of $84 per unit. The maximum output of the factory in normal working conditions is 13 000 units. During the year ended 31 December 2016 the company made and sold 12 000 units. The costs were as follows:

Direct materials	$22 per unit
Direct labour	$14 per unit
Fixed costs	$450 000

(a) State **two** limitations of break-even analysis. [2]

(b) Explain what is meant by the term 'marginal of safety'. [2]

(c) Calculate for the year ended 31 December 2016:

    (i) the break-even point [4]

    (ii) the margin of safety. [1]

Additional information

The company had a target profit of $120 000 for the year ended 31 December 2016.

(d) Calculate the number of units the company needed to sell in order to achieve its target profit. [3]

Additional information

The directors have been concerned by a fall of demand for their product in recent years caused by increasing competition. They have been considering the following alternative courses of action.

Option 1

Reducing the selling price of the product by 12.5 per cent on the current price. The directors expect this would increase demand for the product by 10 per cent. Overtime at time and half would be required for any production outside of normal working hours.

Option 2

Invest in new machinery which would increase current factory capacity by 5 per cent, improve the quality of the product and reduce the labour hours required by 10 per cent due to greater efficiency. The directors believe the factory could operate at maximum capacity and that all production could be sold at the current price. However, fixed costs would increase by $50 000 per annum due to additional depreciation costs and finance charges on the loan which would be required.

Option 3

Switch to producing a similar product. The new product would be superior to the one currently produced and could be sold for a price 20 per cent higher. The new product would require the use of more materials of a higher grade. Each unit would require 10 per cent more materials and these materials will cost 10 per cent more. Labour costs are expected to be unchanged, but existing machinery would require alteration resulting in additional depreciation charges of $20 000 per annum. The directors expect demand will be 10 500 units per annum.

(e) **(i)** Evaluate the options available to the company taking account of financial and non-financial factors. Support your answers with calculations. [15 marks]

    (ii) Recommend which option the directors should choose. Justify your answer. [3 marks]

[Total: 30]

Paper 3 structured questions

Paper 3 consists of four structured questions on financial accounting, and two structured questions on cost and management accounting. This paper tests the additional content for the A Level syllabus, but also requires a knowledge and understanding of the AS Level syllabus content. The paper is worth a total of 100 marks.

1 C Limited is a manufacturer. The following information is available at 31 December 2016.

	$000		$000
Carriage inwards on raw materials	11	Plant and machinery:	
Direct labour	114	cost	345
Distribution expenses	29	provision for depreciation	135
Inventories at 1 January 2016:		Power	50
finished goods	36	Purchases of finished goods	50
raw materials	47	Purchases of raw materials	224
work-in-progress	13	Rent and insurance	62
Inventories at 31 December 2016:		Returns inwards	9
finished goods	33	Returns outwards	12
raw materials	57	Revenue	695
work-in-progress	17	Wages and salaries	70
Office furniture and equipment:			
cost	40		
provision for depreciation	12		

Additional information

1 Power costs should be divided between the factory and office in the ratio 4:1. Eighty per cent of factory power is a direct cost and the rest is an indirect cost.

2 At 31 December rent due but unpaid was $5000, and insurance was prepaid $7000. Rent and insurance should be divided between the factory and office in the ratio 3:1.

3 40 per cent of wages and salaries is an indirect factory cost, the rest applies to office staff.

4 Depreciation policies:

- Machinery and plant: 20 per cent per annum using the reducing balance method.

- Office furniture and equipment: 15 per cent annum using the straight-line method.

Required

(a) Explain how direct costs differ from indirect costs. [3]

(b) Prepare the manufacturing account for the year ended 31 December 2016. [10]

(c) Prepare the income statement for the year ended 31 December 2016. [4]

Additional information

The Finance Director is proposing to introduce a system which identifies manufacturing profit. He says this will result in inventories of finished goods being valued at transfer price.

(d) Identify the two concepts which should be applied when finished goods are valued at transfer price. [2]

(e) Discuss the Finance Director's proposal to identify manufacturing profit. [6]

[Total: 25]

2 R Limited is a company formed to take over the partnership business of Ayesha and Bashir who have been sharing profits and losses in the ratio 3:2.

The takeover took place on 1 December 2016. The partnership statement of financial position at close of business on 30 November 2016 was as follows:

Assets	$	$
Non-current assets (at net book value)		
Property	90 000	
Motor vehicles	37 400	
Furniture and equipment	21 800	
		149 200
Current assets		
Inventory	9 250	
Trade receivables	8 460	
Cash and cash equivalents	3 290	
		21 000
Total assets		170 200
Capital and liabilities		
Capital accounts		
Ayesha	100 000	
Bashir	60 000	
		160 000
Current accounts		
Ayesha	(3 600)	
Bashir	1 900	
		(1 700)
Current liabilities		
Trade payables		11 900
Total capital and liabilities		170 200

The agreed terms of the takeover were as follows:

• Bashir was to keep one of the motor vehicles with a net book value of $25 000 for her private use at a value of $21 500.

• R Limited was to take over all the remaining assets and liabilities, except part of the cash and cash equivalents amounting to $2500. Certain assets were to be revalued as follows:

	$
Property	120 000
Remaining motor vehicle	11 000
Furniture and equipment	17 500
Inventory	8 400
Trade receivables	7 900

The cost of dissolving the partnership was $3500.

The purchase consideration was $180 000 to be discharged by the issue of 120 000 $1 ordinary shares to the partners in proportion to the balances on their capitals accounts on 30 November 2016.

The partners were to settle the balances remaining on their capital accounts by making payments into or from the partnership bank account from their private resources.

Required

(a) Explain **two** benefits which may have encouraged Ayesha and Bashir to convert their business to a limited company. [4]

(b) Calculate the realisation profit or loss made on the dissolution of the partnership. [5]

(c) Prepare the partners' capital accounts showing the final settlement on the dissolution of the partnership. [7]

(d) Prepare the statement of financial position of R Limited immediately after the takeover of the partnership on 1 December 2016. [5]

Additional information

The partners had forecast a profit for the year ended 30 November 2017 of $60 000. The deed of partnership had required interest on capitals at 10 per cent per annum to be allowed before sharing remaining profits and losses.

The partners are assuming that R Limited will achieve the same forecast profit and all profits will be distributed. They do not anticipate making any further issue of shares in the first year of business.

(e) Assess the effect of the change from a partnership to a limited company on each partner in the year ending 30 November 2017. [4]

[**Total: 25**]

3 The treasurer of the Riverside Community Club has provided the following information for the year ended 30 September 2016.

The club's assets and liabilities including the following:

	At 1 October 2015 $	At 30 September 2016 $
Furniture and equipment at net book value	43 450	44 100
Café inventory	1 920	1 830
Trade payables for café supplies	5 130	4 530
Subscriptions received in advance	900	1 200
Subscriptions owing	2 100	1 400
Life membership fund	15 600	?
Interest free loans from members	10 000	?
Clubhouse rent prepaid	720	
Clubhouse rent owing		640

Receipts and payments account
for the year ended 30 September 2016

	$	$
Opening balance overdrawn		(3 930)
Receipts		
Café takings	22 480	
Donations	1 200	
Members' subscriptions	15 600	
Life membership fund	4 800	
Loans from members	3 500	
Sale of unused furniture	220	
		47 800
Payments		
Suppliers of café refreshments	13 870	
Clubhouse rent	9 260	
Additional furniture	4 840	
Refund of members' loans	1 200	
Café staff wages	7 350	
Administration expenses	2 610	
		(39 130)
Closing balance		4 740

The club's policy is to regard 15 per cent of the life membership fund as income for the year under review.

The unused furniture sold during the year had a net book value of $80 at 1 October 2015.

Required

(a) Calculate the club's accumulated fund at 1 October 2015. [2]

(b) Prepare the club's income and expenditure account for the year ended 30 September 2016. [10]

(c) Prepare the club's statement of financial position at 30 September 2016. [5]

Additional information

The club's committee has considered a proposal to close the café as profits have been falling over recent years. A member has suggested that the space could be used for other profit-making activities and that the members' loans should be used to upgrade this area of the clubhouse to make it more attractive.

(d) Advise the committee on the proposal. Justify your advice. [8]

[Total: 25]

4 Y Limited manufacture three products: 'jed', 'ked' and 'led'. The company uses activity based costing and has identified the following cost pools and cost drivers.

Cost pools	Cost drivers
Machining department	Number of machine set ups required to produce each product.
Inspections	Number of inspections required during the manufacture of each product.

The following forecasts have been made for August 2017:

Production and sales

Jed	2 000 units
Ked	1 000 units
Led	3 000 units

Overheads

Cost pool	Total overheads $	Information about each product		
		Jed per unit	Ked per unit	Led per unit
Machining	27 500	1 set up	2 set ups	2 set ups
Inspections	9 900	2 inspections	2 inspections	3 inspections

Each 'jed' requires following direct costs:

Direct materials	3.2 kilos at $1.90 per kilo
Direct labour	3.6 hours at $9.80 per hour

The selling price of a 'jed' is calculated to achieve a gross margin of 20 per cent.

Required

(a) Explain the meaning of the term 'cost driver'. [2]

(b) Calculate the overhead allocation rate for each cost pool. [4]

(c) Calculate the total overheads allocated to each product. [7]

(d) Calculate the selling price of a 'jed'. [6]

Additional information

The directors have been informed that a rival company will market a product similar to the 'jed'. In order to complete Y Limited would have to reduce its profit margins on this product to very low levels. The directors are proposing to cease production of 'jeds', and to switch production capacity to produce more 'keds' and 'leds'.

(e) Discuss the proposal to cease production of 'jeds' and advise the directors on the best course of action. Justify your advice. [6]

[Total: 25]

Index